Precancer

The Beginning and the End of Cancer

Related Titles from Jones and Bartlett Publishers

Apoptosis, Genomic Integrity, and Cancer
Julien L. Van Lancker

Bioimaging: Current Concepts in Light and Electron Microscopy
Douglas E. Chandler & Robert W. Roberson

Biomedical Graduate School: A Planning Guide to the Admissions Process
David J. McKean & Ted R. Johnson

The Cancer Book
Geoffrey M. Cooper

CELLS
Benjamin Lewin, Lynne Cassimeris, Vishwanath R. Lingappa, & George Plopper

Electron Microscopy, Second Edition
John J. Bozzola & Lonnie D. Russell

Elements of Human Cancer
Geoffrey M. Cooper

Fundamentals of Cancer Epidemiology, Second Edition
Phillip C. Nasca & Harris Pastides

Genomic and Molecular Neuro-Oncology
Wei Zhang & Gregory N. Fuller, eds.

Glioblastoma Multiforme
James Markert, Vincent T. DeVita Jr., Steven A. Rosenberg, & Samuel Hellman

Handbook of Cancer Risk Assessment and Prevention
Graham A. Colditz & Cynthia J. Stein

How Cancer Works
Lauren Sompayrac

Human Embryonic Stem Cells, Second Edition
Ann A. Kiessling & Scott C. Anderson

Malignant Liver Tumors: Current and Emerging Therapies, Second Edition
Pierre Clavien, Yuman Fong, H. Kim Lyerly, Michael A. Morse, & Alan P. Venook

Molecular Oncology of Breast Cancer
Jeffrey S. Ross & Gabriel N. Hortobagyi

Molecular Oncology of Prostate Cancer
Jeffrey S. Ross & Christopher S. Foster

Neoplasms: Principles of Development and Diversity
Jules J. Berman

Oncogenes, Second Edition
Geoffrey M. Cooper

Pediatric Stem Cell Transplantation
Paulette Mehta, ed.

Physicians' Cancer Chemotherapy Drug Manual 2009
Edward Chu

Protein Microarrays
Mark Schena, ed.

Precancer

The Beginning and the End of Cancer

Jules J. Berman, MD, PhD
with G. William Moore, MD, PhD

JONES AND BARTLETT PUBLISHERS
Sudbury, Massachusetts
BOSTON TORONTO LONDON SINGAPORE

World Headquarters
Jones and Bartlett Publishers
40 Tall Pine Drive
Sudbury, MA 01776
978-443-5000
info@jbpub.com
www.jbpub.com

Jones and Bartlett Publishers
Canada
6339 Ormindale Way
Mississauga, Ontario L5V 1J2
Canada

Jones and Bartlett Publishers
International
Barb House, Barb Mews
London W6 7PA
United Kingdom

Jones and Bartlett's books and products are available through most bookstores and online booksellers. To contact Jones and Bartlett Publishers directly, call 800-832-0034, fax 978-443-8000, or visit our website, www.jbpub.com.

Substantial discounts on bulk quantities of Jones and Bartlett's publications are available to corporations, professional associations, and other qualified organizations. For details and specific discount information, contact the special sales department at Jones and Bartlett via the above contact information or send an email to specialsales@jbpub.com.

Production Credits
Chief Executive Officer: Clayton Jones
Chief Operating Officer: Don W. Jones, Jr.
President, Higher Education and Professional Publishing: Robert W. Holland, Jr.
V.P., Sales: William J. Kane
V.P., Design and Production: Anne Spencer
V.P., Manufacturing and Inventory Control: Therese Connell
Publisher, Higher Education: Cathleen Sether
Acquisitions Editor: Molly Steinbach
Associate Editor: Katharine F. Theroux
Editorial Assistant: Caroline Perry
Production Manager: Louis C. Bruno, Jr.
Senior Marketing Manager: Andrea DeFronzo
Text Design: Anne Spencer
Cover Design: Scott Moden
Illustrations: Elizabeth Morales
Permissions Coordinator: Kesel Wilson
Composition: M&M Composition, LLC
Cover Image: Courtesy of G. William Moore, MD, PhD, Pathology and Laboratory Medicine Service, Baltimore VA Medical Center, Baltimore, MD
Printing and Binding: Malloy, Inc.
Cover Printing: Malloy, Inc.

Library of Congress Cataloging-in-Publication Data
Berman, Jules J.
Precancer : the beginning and the end of cancer / Jules J. Berman. — 1st ed.
p. ; cm.
Includes bibliographical references and index.
ISBN 978-0-7637-7784-5 (alk. paper)
1. Precancerous conditions. 2. Cancer—Treatment. I. Title.
[DNLM: 1. Precancerous Conditions—pathology—Popular Works. 2. Clinical Trials as Topic—statistics & numerical data—Popular Works. 3. Neoplasms—prevention & control—Popular Works. QZ 201 B516p 2010]
RC268.5.B48 2010
616.99'4071—dc22 2009023830

6048

Printed in the United States of America

13 12 11 10 09 10 9 8 7 6 5 4 3 2 1

Brief Contents

Contents

Part III Eradication of Cancer By Treatment of Precancers 89

Preface

The beginnings and endings of all human undertakings are untidy.

—John Galsworthy

When people think about cancer, they envision large, aggressive tumors that invade through tissues and metastasize to distant parts of the body. Most people are unaware that almost every cancer passes through a small precancer phase, during which it cannot metastasize or invade.

Medical science has been largely unsuccessful at curing advanced cancers. This book explains that recent advances in our understanding of carcinogenesis (the biological process leading to the development of cancer) have helped us to develop strategies to prevent, diagnose, and treat precancers, the relatively innocuous lesions that precede cancers. The number of people who die each year from cancer would drop precipitously if cancer researchers focused their efforts on curing precancers.

You may be wondering why you have heard about cancer all your life and have seldom (if ever) heard the term *precancer*. Maybe you have heard of some types of precancers without having heard the collective term for these lesions.

Have you ever been examined by a dermatologist who found a small roughened or reddened area on your face, neck, arms, or back and who told you that the lesion was an actinic keratosis? Actinic keratoses are the most common precancers occurring in humans. Have any of your women friends been diagnosed with cervical intraepithelial neoplasia (CIN) of the uterine cervix? CIN is the most common precancer occurring in the cervix. Although the term *precancer* is seldom used in magazine articles or in your doctor's office, precancers are the most commonly encountered growths in the field of cancer medicine.

Precancer treatment is responsible for the most dramatic reduction in cancer deaths attributable to any medical intervention. The current generation of Americans may forget that cancer of the uterine cervix was once the leading cause of cancer deaths in women. Every country that has instituted population-wide cervical precancer screening has had a 70% to 90% reduction in uterine cancer mortality. In the United States alone, cervical precancer treatment saves over 60,000 lives every year. No effort aimed at treating invasive cancers has provided an equivalent reduction in cancer deaths.

Have you, or any of your friends and relatives, been diagnosed with a colon adenoma, also known as benign adenomatous polyp of the colon? Colon adenomas are one type of very common precancer. Most colon adenomas can be excised (removed) during a colonoscopy. A recent report issued by the National Cancer Institute concluded that "evidence suggests that 90% or more of colorectal cancer deaths could be prevented if precancerous polyps were detected with routine screening and removed at an early stage (1)." In 2008, 49,960 Americans died of colon or rectal cancer (2). Though the prediction of a 90% reduction in colon cancer deaths with routine screening has been adjusted downwards (3), it is still worth noting that a 90% reduction in the colon cancer death rate would save 45,000 U.S. lives each year.

The purpose of this book is to show that we currently have the medical knowledge needed to prevent most common cancers from developing, using available public health measures that prevent or treat precancers. Because precancers are easier to cure than cancers, we can expect that new agents designed to treat precancers will prevent the occurrence of most human cancers in the not-too-distant future. Everyone involved in healthcare can benefit from a deeper understanding of precancers.

Organization of *Precancer*

You will not fully appreciate the importance of the precancers until you have a full understanding of the limitations of past and current approaches to cancer treatment. The book is divided into three parts: Part I, Limitations of Current Approaches to Cancer Treatment; Part II, Precancer Pathology and Biology; and Part III, The Eradication of Cancer by Treatment of Precancers.

Part I. Limitations of Current Approaches to Cancer Treatment

In the past decade, we have learned a great deal about the genetics of cancer. Unfortunately, the most important news in the field of cancer genetics is bleak: cancers are much more complex than we had feared. The most common cancers of humans have many thousands of genetic alterations, and the causes of these genetic alterations cannot be traced to any single initiating factor. Because cancers are genetically complex, it is unlikely that we will cure the common cancers of humans by developing drugs that correct any single mutation in the cancer genome.

Chapters 1 and 2 explain that the U.S. cancer death rate has been virtually constant through the past half-century. The common cancers that were killing us in the mid–twentieth century are the same cancers that are killing us today, in about the same numbers. Chapter 3 explains why curing the common advanced cancers has been so difficult, and why curing several of the rarest types of cancer has been possible. The biological distinction between common cancers and rare cancers is a key concept that will help us understand the role of precancers in cancer eradication.

Part II. Precancer Pathology and Biology

In Chapters 4 and 5, the pathological and biological properties of precancers are discussed. Every reader, regardless of his or her medical background, can benefit from an explanation of the fundamental changes in cells that characterize the precancers and that distinguish precancers from normal tissues and from cancerous tissues.

Sometimes, the easiest way to understand complex systems is to study the simple systems from which they develop. In Chapter 6, we will see that by studying precancers, we can examine questions that cannot be answered by studying advanced cancers.

Part III. Eradication of Cancer by Treatment of Precancers

In Chapter 7, we show that many treatments have been proved effective against precancers, and we review some of the newest approaches to precancer treatment currently under development. It is often said that science is 10% research and 90% politics. A scientific field cannot possibly advance without the willing participation of scientists, the enthusiasm of the public, and the financial backing of funding agencies. Chapter 8 delves into some of the entrenched misconceptions about precancers. Chapter 9 lays out various scenarios for the impact of cancer in our futures. The book urges the cancer care community to move forward, now, with a vigorous campaign focused on the precancers.

Appendices

Following the chapters are three appendices. Appendix 1 assembles statistical data used in the text and provides sources for each value. Appendix 2 explains precancer terminology and organizes the generally recognized precancer lesions under an anatomic schema. Appendix 3 describes the Developmental Lineage Classification and Taxonomy of Cancer, which subsumes precancer terminology. This free and open-source document is the largest extant terminology of neoplasms and been described in detail in *Neoplasms: Principles of Development and Diversity*, also published by Jones and Bartlett Publishers, LLC. The Developmental Lineage Classification and Taxonomy of Cancer is available on the author's website, http://www.julesberman.info/devclass.htm.

References

Because the book makes some criticism of the current direction of cancer research in the United States and elsewhere, providing primary references that support every assertion is important. Over 100 references are cited in the text. Many of these references are free, publicly available documents, and their Web addresses are included wherever possible.

Glossary

Some specialized medical terminology appears in the text, but most technical terms are explained when they first appear. In some cases, full explanations of terms encountered would disrupt the narrative, so the book's glossary provides a definition, along with an explanation of the term's biomedical importance. First occurrences of technical terms in the text that appear in the Glossary are in bold.

Who Should Read This Book?

Preventing and treating precancers is the most effective way to reduce the number of cancers. After reading *Precancer: The Beginning and the End of Cancer,* you will have a good understanding of precancer biology and the various methods that can prevent, detect, or eliminate precancers. This book is written primarily for students and professionals who can implement the strategies discussed in the book: public health professionals, clinicians, nurses, hospital administrators, government healthcare planners, and health insurance managers. Motivated laypersons who have an interest in cancer prevention and treatment will find that the text can be understood with the aid of the book's extensive Glossary.

Jules J. Berman

Author Biographies

Jules Berman, MD, PhD, has studied cancer for the past 36 years. After earning two bachelor of science degrees (mathematics and earth sciences from MIT), he entered the graduate program in pathology at Temple University, where he began his thesis work in the Fels Cancer Research Institute. He spent the final year of his graduate studies at the Naylor Dana Institute of the American Health Foundation in Valhalla, NY, before beginning his postdoctoral studies in the Perinatal Carcinogenesis Section of the Laboratory of Experimental Pathology at the U.S. National Cancer Institute, Bethesda, MD. He earned his medical degree from the University of Miami, and he completed his residency in anatomic pathology at the George Washington University Medical Center in Washington, DC. He became board certified in anatomic pathology and in cytopathology and served as the Chief of Anatomic Pathology, Surgical Pathology, and Cytopathology at the Veterans Affairs Medical Center in Baltimore, MD. Dr. Berman has held adjunct appointments at the University of Maryland Medical Center and at the Johns Hopkins Medical Institutions, Baltimore, MD. In 1998, he was appointed Program Director and Medical Officer in the Cancer Diagnosis Program at the U.S. National Cancer Institute. He served as President of the Association for Pathology Informatics in 2006. Over the course of his career, he has first-authored more than 100 publications and has contributed to hundreds of scientific articles. Dr. Berman has written six books—most recently, *Neoplasms: Principles of Development and Diversity.*

G. William Moore, MD, PhD, earned a Bachelor of Science, majoring in cellular biology, from the University of Michigan, Ann Arbor, a PhD in biomathematics from North Carolina State University, Raleigh, and an MD at Wayne State University, Detroit, MI. He worked as a research assistant at the University of Freiburg Pathology Institute, Freiburg, Germany. He completed his residency in anatomic pathology at the Johns Hopkins Hospital, Baltimore, MD, and is board certified in anatomic pathology. Dr. Moore has appointments as Assistant Professor of Pathology at the Johns Hopkins

Medical Institutions and Associate Professor of Pathology at the University of Maryland, in Baltimore. He serves as Staff Pathologist and Director of Autopsy Pathology at the Veterans Affairs Medical Center, in Baltimore. In 2007, he received the Honorary Fellow Award of the Association for Pathology Informatics. Over the past 40 years, he has co-authored over 180 publications in peer-reviewed medical journals.

PART

Limitations of Current Approaches to Cancer Treatment

CHAPTER

The Cancer Burden

1

All interest in disease and death is only another expression of interest in life.

—Thomas Mann

1.1 Background

The estimate for the total number of deaths worldwide from cancer in 2001 was 7,021,000 (4). In the United States, cancer currently accounts for 23% of all deaths, up from 19% in 1975 (Figure 1.1). Today, the U.S. lifetime risk of developing cancer is 41% (5).

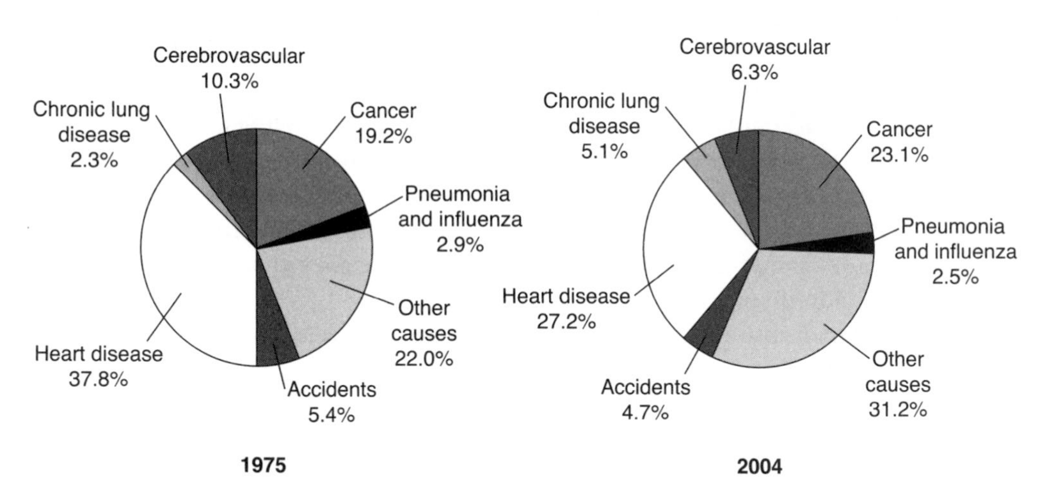

FIGURE 1.1 Leading causes of death in the United States as a percent of all causes of death, in 1975 and 2004, from the National Center for Health Statistics public use data file (6). Most major causes of death have dropped between 1975 and 2004, while the percentage of deaths resulting from cancer has increased.

The purpose of this chapter is to show that, despite advances in cancer research and cancer treatment, the worldwide burden of cancer has been steadily increasing.

1.2 Historical Cancer Death Rates

Is cancer a man-made plague? Has the incidence of cancer increased as humans have populated, industrialized, and polluted the planet? We know very little about the occurrence of cancers in ancient humans. Examination of mummified skeletons indicates that either **primary** or metastatic **tumors** have occurred in bones, but we know nothing about the range of tumors that occurred in ancient man or of the overall incidence of cancers during human prehistory.

In the United States in 1913, the American Society for the Control of Cancer was established. This organization was later renamed the American Cancer Society. One of its first actions was to evaluate a widely held perception that the incidence of cancer was increasing rapidly. Frederick L. Hoffman, a statistician and chair of the Committee of Statistics for the Society, collected all available data on cancer incidence and **mortality** and, in 1915, published a fascinating book *The Mortality of Cancer Throughout the World* (7).

He wrote, "Probably the very earliest cancer mortality statistics are contained in the *Collection of the Yearly Bills of Mortality from 1657 to 1758 Inclusive, Together with Several Other Bills of an Earlier Date,* published in London, England in 1759. The returns are limited to the city of London. The incidence of cancer is specifically enumerated for each year; however, the use of the term *cancer* includes gangrene and fistula. Excluding the years during which plague was epidemic, between 1651 and 1758, out of 1,980,037 deaths from all causes, 5,123, or 0.26%, were from cancer, including gangrene and fistula. The proportion was highest between 1651 and 1664, or 0.34%, and lowest between 1741 and 1758, or 0.20%.

Today, cancer accounts for about 23% of deaths in the United States and this represents an 88-fold increase compared to the 0.26% figure reported for seventeenth century London. Is this number reliable? Not very. Changes in life expectancy and accuracy of diagnosis have improved in the interim centuries, but the stark contrast in reported numbers raises suspicions that cancer has been a growing problem in the past several hundred years.

In the late nineteenth century, cancer statistics were collected in the United States. Hoffman writes,

> *Another illustration is the experience of the State of Massachusetts. In 1871 the recorded cancer death rate was 36.9 per 100,000 of population; by 1881 the rate had increased to 52.3; by 1891 to 60.9; by 1901 to 73.1, and by 1911 to 92.6....*
>
> *Boston has for many years been one of the medical centers not only of the United States but of the world. There are no reasons for believing that medical*

diagnosis was so crude or imperfectly developed in 1871 that one out of every two deaths from cancer should have been erroneously diagnosed or wrongfully classified under some other disease. Nor is there any evidence to substantiate the point of view that the age distribution of Massachusetts has undergone such profound changes as to account for the higher frequency of cancer at the present time. In 1880 the proportion of population ages 65 and over in Massachusetts was 5.4 per cent; in 1900 it was 5.1 per cent; in 1910 it was 5.2 per cent. From a practical point of view in statistical analysis, these changes in the age distribution can have been of only slight effect on the cancer death rate (7).

Taking Hoffman's data at face value, there has been about a five-fold increase in the observed U.S. cancer death rate from the mid-nineteenth century to the mid-twentieth century. (List 1.2.1).

In 1915, when Hoffman assembled his data, evidence suggested that a synchronous increase in the death rate from cancer was occurring throughout, in his words, the civilized world.

Results of the present investigation, which prove that within less than forty years the rate of mortality from cancer has practically doubled and that the actual number of deaths from cancer in the civilized portion of the world for which reasonably trustworthy data are available exceeds 500,000 per annum.... Combining the returns for the United Kingdom, Norway, Holland, Prussia, Baden, Switzerland, Austria, the cities of Denmark, the Commonwealth of Australia, and the Dominion of New Zealand, it appears that these countries in 1881 had an aggregate population of 98,380,000 and 44,047 deaths from cancer, equivalent to a rate of 44.8 per 100,000 of population; by 1891 the rate had increased to 59.6, by 1901 to 76.3 and by 1911 to 90.4. Thus, during thirty years the cancer death rate in these countries, which are typical of the civilized portion of the world, has more than doubled, or, to be exact, the rate for 1911 was 101.8 per cent in excess of the rate prevailing in 1881 (7).

Nearly 100 years ago, there was a perception that the **incidence** of cancer was increasing in the United States. This perception was confirmed using the best available

List 1.2.1 Historical incidence of cancer (per 100,000 population).

U.S. Cancer Death rate, 1871 (not age-adjusted): 36.9 (7)
U.S. Cancer Death rate, 1881 (not age-adjusted): 52.3 (7)
U.S. Cancer Death rate, 1891 (not age-adjusted): 60.9 (7)
U.S. Cancer Death rate, 1901 (not age-adjusted): 73.1 (7)
U.S. Cancer Death rate, 1911 (not age-adjusted): 92.6 (7)
U.S. Cancer Death rate, 1950 (age-adjusted): 195.4 (8)

evidence at that time. Using the same data, the incidence of cancer deaths in 1911 was about half of what it is today.

In 1971, President Richard M. Nixon signed the National Cancer Act into law, marking the year that the United States launched its **war on cancer**. For the next two decades, the U.S. cancer death rate rose steadily. Then in 1991, the U.S. cancer death rate began to decline, incrementally but continuously (Figure 1.2). Is this a sign that the war on cancer is being won and that cancer treatment has improved to the point that we can find clear statistical evidence that cancer can be defeated?

One may be tempted to conclude that 1991 marked the beginning of victory in our war against cancer, and that the steady, incremental declines in U.S. cancer death rates will continue in future decades, until cancer is fully eradicated.

This is simply not the case. Examination of the **National Cancer Institute's (NCI) Surveillance, Epidemiology and End Results (SEER)** graph clearly shows that the small decline in the cancer rate since 1991 is counter-balanced by a small rise in the rate of cancer deaths between 1975 and 1991. Today, we have about the same cancer death rate as we had in 1975. Cancer researchers did not accept blame for the rise in cancer before 1991. Likewise, cancer researchers should not accept credit for the drop in cancer after 1991. Furthermore, if you look at death rate data extended over a half century of observation, the small decrease in cancer deaths since 1991 is imperceptible (Figure 1.3) and far less impressive than gains made in other diseases (9).

Age-adjustment is an established procedure, performed by statisticians who are comparing disease rates in different populations (or in one general population at different times in history). Age-adjustment normalizes changes in disease rates caused by shifts in the ages of people in the population. If a disease exclusively strikes people over the age of 60 and the population has a large increase in the over-60 population, then it would necessarily have an increase in the incidence of disease on this basis, even when there are no medical or biological causes for the increase. To compensate for this effect, statisticians "adjust" the data to normalize the distribution of ages against a standard age distribution (currently, the year 2000 age distribution determined by the U.S. Census Bureau).

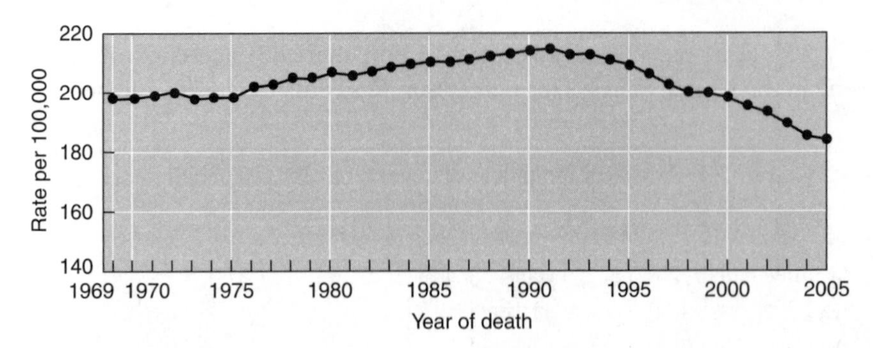

FIGURE 1.2 Age-adjusted total U.S. mortality rates for all cancer sites, male and female, all ages, and all races for 1969–2005, age-adjusted to the year 2000 U.S. standard population. Data provided by the NCI's SEER program (http://seer.cancer.gov/canques/mortality.html).

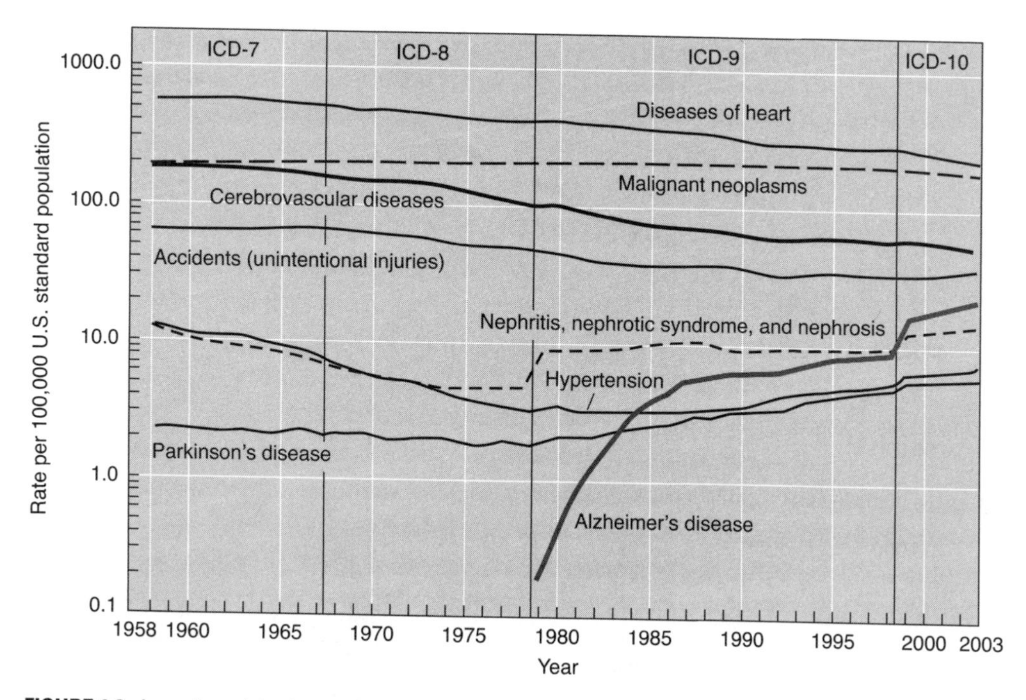

FIGURE 1.3 Age-adjusted death rates for selected leading causes of death in the U.S., 1958–2003. Data from Hoyert DL, Heron MP, Murphy SL, Kung H-C. Final Data for 2003. *National Vital Statistics Report* 54:(13), April 19, 2006.

Age-adjusted death rate is not a suitable measure for the burden of cancer on a population. If you want to know the burden of a disease on a population, you need to use the crude (unadjusted) cancer death rate (the number of people who die from cancer per 100,000 population), and the total number of people who die from cancer (in the U.S. population). These numbers reflect the burden of cancer on healthcare resources and on society. Figure 1.4 is a graph of the crude death rate data for all cancers in the United States between 1969 and 2005. When we examine the unadjusted (crude) cancer

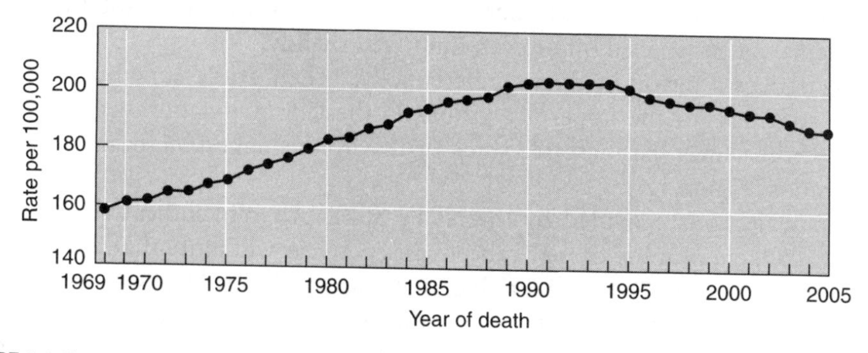

FIGURE 1.4 Unadjusted death rate for cancer in the U.S., 1969–2003. Data provided by the NCI's SEER program (http://seer.cancer.gov/canques/mortality.html).

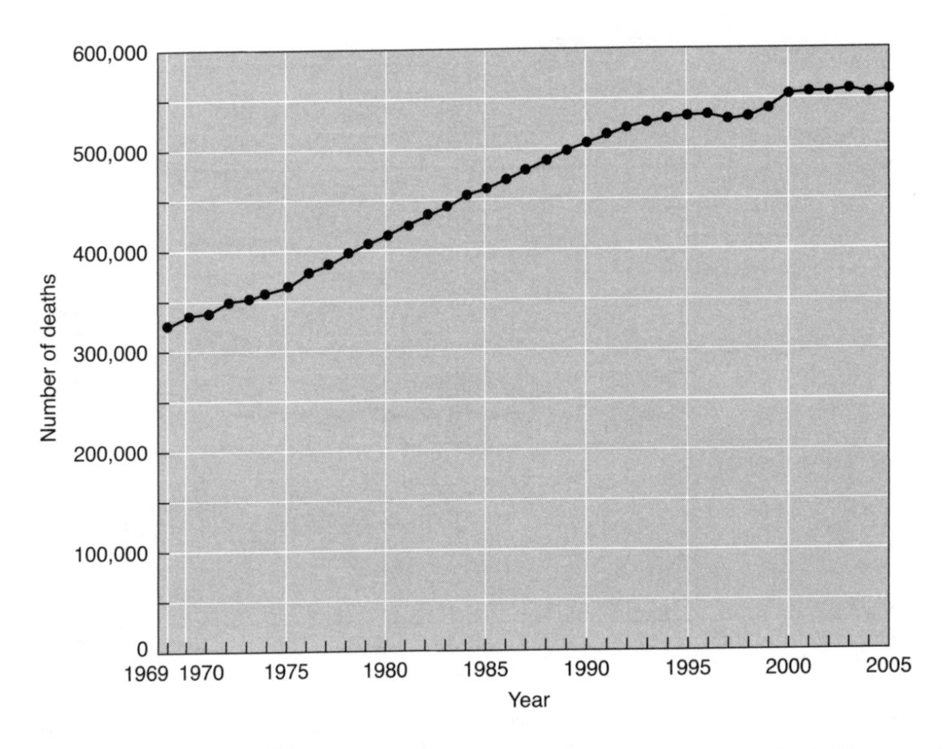

FIGURE 1.5 Total deaths due to cancer in the U.S., 1969–2003. Data provided by the NCI's SEER program (http://seer.cancer.gov/canques/mortality.html).

death rates from 1969 to 2005, we find that that despite the dip in cancer incidence from 1991 to the present, cancer deaths (per 100,000 population) have increased since 1970.

To find the total number of deaths from cancer, you multiply the crude death rate for each year by the population of the United States during that year, obtained from the U.S. Census bureau and expressed as hundreds of thousands. The number of people dying from cancer in the United States just gets higher and higher. Again toward the end of the graph (Figure 1.5), there are a few years in which the total number of deaths seems to have leveled off or even improved slightly.

Cancer strikes older individuals preferentially. When crude (unadjusted) cancer dates are stratified for the upper age groups in the U.S. population, we see that the death rate from cancer has increased since 1970, with the highest increases in the highest age groups (Figure 1.6).

Cancer projections provided by the NCI's SEER program indicate that between 2000 and 2050, the number of new cancer cases per year will more than double, from 1.3 million new cases in 2000 to 2.8 million new cases in 2050 (11). Will the American healthcare system provide adequate care for the increasing number of cancer cases that will occur, in the near future?

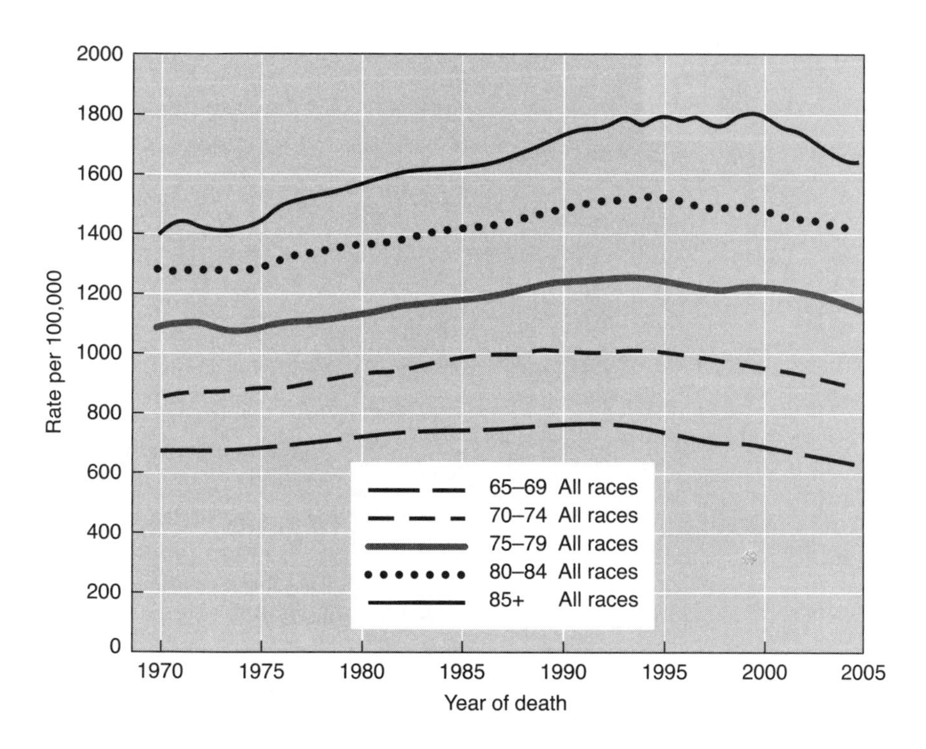

FIGURE 1.6 Crude cancer mortality, all races, male and female, stratified by age ranges in adults over age 60, rates per 100,000 population. Data from Surveillance, Epidemiology, and End Results (SEER) Program (www.seer.cancer.gov). *Mortality, Total U.S. (1969–2005).* National Cancer Institute, DCCPS, Surveillance Research Program, Cancer Statistics Branch, April 2008. Underlying mortality data provided by NCHS (www.cdc.gov/nchs).

1.3 Drop In Cancer Death Rate Since 1991

Let us assume, for the sake of argument, that the steady decline in cancer deaths since 1991 is real and not a statistical aberration. Is the drop in cancer deaths the result of improvements in the treatment of **advanced cancers**? No. The overall drop in cancer deaths is largely the result of drops in incidence of a few types of cancer and from the treatment of precancers and early cancers that have not metastasized prior to their clinical detection. If we look at trends in lung cancer incidence and mortality from lung cancer, we see that the drop in mortality from this cancer is largely the result of smoking cessation in men (Figure 1.7). In addition, the U.S. lung cancer trends between 1975 and 2004 tell a great deal about the small drop in the overall U.S. cancer deaths that has occurred since 1991. First, the drop in cancer deaths in the United States is synchronous with the large, but isolated, drop in lung cancer incidence and deaths in men that began in 1991. Second, the graph indicates that the drop in lung cancer death rates has

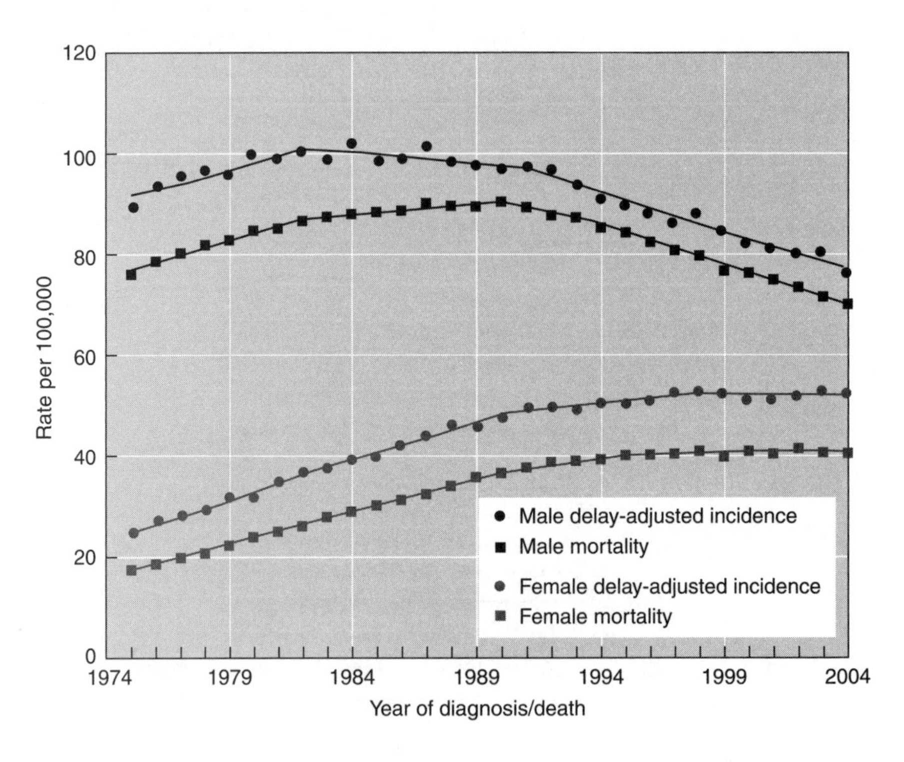

FIGURE 1.7 Age-adjusted incidence and mortality from cancer of the lung and bronchus in the U.S., 1975–2004. Data from SEER. *Cancer Statistics Review 1975–2004.* National Cancer Institute. http://seer.cancer.gov/csr/1975_2004/results_merged/sect_01_overview.pdf, pages 30–31, Table I-4.

not extended to women, whose cigarette-associated peak in lung cancer deaths has not yet occurred. Third, the lines of incidence and mortality for lung cancer are roughly parallel, and this holds true for both men (top lines) and women (bottom lines). This indicates that the lung cancer cure rates have not improved to any noticeable extent during the 30 years of observation. Fourth, the close proximity of the lines for cancer incidence and cancer deaths indicates that the vast majority of patients with lung cancer will die from their disease.

Lung cancer is the leading cause of cancer deaths in the United States. The overall five-year survival for lung cancer is a little over 15%. More than one-third of patients with lung cancer will be diagnosed with Stage IV disease (12). Stage IV lung cancers have spread to lymph nodes and to another lobe of the lungs or to other parts of the body. The chances of a person with Stage IV lung cancer surviving five years is under 2% (Figure 1.8). The chances of being completely cured of lung cancer is even smaller. Despite decades of cancer research and many billions of dollars spent, the cure rate for patients with advanced lung cancer is very low. The only effective means of reducing lung cancer deaths is through prevention.

Stage	Male and Female Cases	Male and Female Percent	Male and Female 5-Year Relative Survival Percent	Male Cases	Male Percent	Male 5-Year Relative Survival Percent	Female Cases	Female Percent	Female 5-Year Relative Survival Percent
All stages	201,067	100.0	15.5	117,472	100.0	13.6	83,595	100.0	18.0
I	26,879	13.4	56.9	14,598	12.4	53.5	12,281	14.7	60.8
II	5,635	2.8	33.7	3,402	2.9	32.4	2,233	2.7	35.7
III	50,254	25.0	9.4	29,863	25.4	9.0	20,391	24.4	10.1
IV	75,057	37.3	1.8	44,783	38.1	1.6	30,274	36.2	2.2
Unknown	43,242	21.5	18.0	24,826	21.1	15.0	18,416	22.0	21.9

FIGURE 1.8 Stage-specific survival for lung cancer. Data from Lynn A. Gloeckler Ries and Milton P. Eisner. Chapter 9, Cancer of the Lung. In, Ries LAG, Young JL, Keel GE, Eisner MP, Lin YD, Horner M-J (eds). *SEER Survival Monograph: Cancer Survival Among Adults: U.S. SEER Program, 1988–2001, Patient and Tumor Characteristics.* National Cancer Institute, SEER Program, NIH Pub. No. 07-6215, Bethesda, MD, 2007.

Is the drop in lung cancer mortality in men since 1991 sufficient to account for the total drop in the U.S. cancer death rate from all sites combined? Not quite. The NCI's SEER program provides annual age-adjusted cancer mortality data for the common cancer sites and for all sites combined (Figure 1.9). A reduction in deaths from lung, colon, and prostate cancers account for the bulk of the overall drop in cancer deaths following 1991 (13) (List 1.3.1).

The total drop in cancer deaths in men from 1991 through 2004 was 50.6 deaths per 100,000. Lung, colon, and prostate contributed drops of 19.6, 8.0, and 13.8 deaths per 100,000 and a combined drop of 41.4 deaths per 100,000. For men, therefore, over 80% of the decrease in cancer mortality in the years 1991 to 2004 is accounted for by synchronous drops in mortality from lung, **colorectal**, and prostate cancer. In each case, the decreases in mortality occurred with tumors that were prevented (smoking cessation preventing lung cancer), with precancers that were detected and excised (**colonoscopic** resection of adenomatous **polyps** preventing colon cancer), or with cancers that were

List 1.3.1 Age-adjusted SEER death rates in U.S. men (deaths per 100,000 men)

Year	All sites	Lung	Colon	Prostate
1991	279.1	89.9	29.6	39.3
2001	244.1	75.2	24.2	29.2
2002	240.2	73.5	23.8	28.2
2003	234.1	71.9	23.0	26.6
2004	228.5	70.3	21.6	25.5

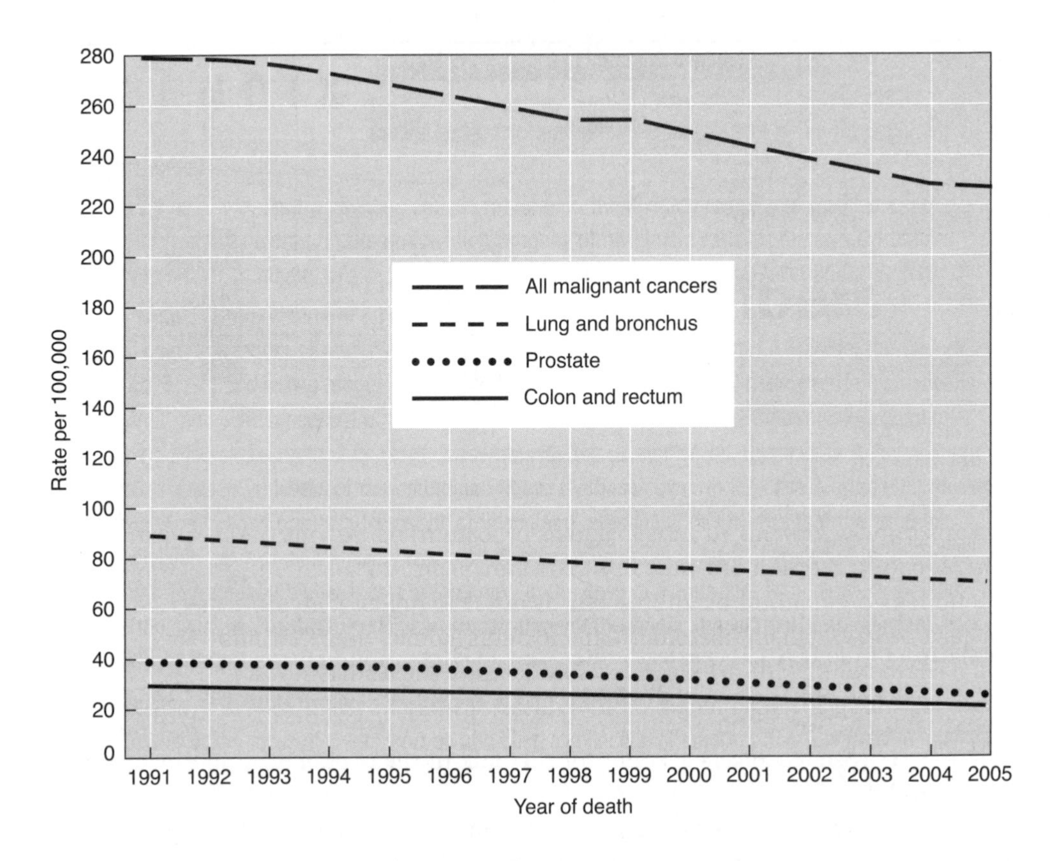

FIGURE 1.9 Age-adjusted SEER death rates in U.S. men for all sites, lung, colon, and prostate cancers for years 1991–2005. Data from Surveillance, Epidemiology, and End Results (SEER) Program (www.seer.cancer.gov). *Mortality, Total U.S. (1969–2005).* National Cancer Institute, DCCPS, Surveillance Research Program, Cancer Statistics Branch, April 2008. Underlying mortality data provided by NCHS (www.cdc.gov/nchs).

treated at an early stage. This indicates that preventive and early intervention measures have been responsible for most of the reduction of cancer deaths in men occurring since 1991.

Chapter 1 Summary

Historical data suggest that the death rate from cancer has been rising throughout the twentieth century and that the burden of new cancer cases will rise steadily during the twenty-first century.

Advanced cancers are those that have metastasized to sites outside the growth zone of the primary cancer. Advanced common cancers account for most deaths due to cancer. Currently, we have no effective methods that can cure the common, advanced cancers that occur in humans. By and large, the same common advanced cancers that were responsible for the greatest numbers of deaths in 1950 are the same cancers killing us today and at about the same rates (14, 15).

In the United States, the total number of cancer deaths has been increasing for at least the past half-century. The largest reduction in the overall U.S. cancer death rate since 1991 is accounted for by the reduction in lung cancer deaths. The drop in lung cancer deaths in men is the result of smoking cessation. The relatively modest reduction in the overall number of U.S. cancer deaths since 1991 has resulted from preventive measures, measures that treat precancer, or measures that treat early cancers (that have not **metastasized**). Improvements in the treatment of advanced common cancers in adults have had a negligible effect on cancer mortality in the U.S. population.

CHAPTER

2 Illusion of Cancer Survival

Deep doubts, deep wisdom; small doubts, little wisdom.

—Chinese proverb

2.1 Background

One cannot pick up a newspaper these days without reading an article proclaiming progress in the field of cancer research. Here is an example, taken from an article posted on the MedicineNet site (16). The lead-off text is

> *Statistics (released in 1997) show that cancer patients are living longer and even "beating" the disease. Information released at an AMA sponsored conference for science writers, showed that the death rate from the dreaded disease has decreased by three percent in the last few years. In the 1940s only one patient in four survived on the average. By the 1960s, that figure was up to one in three, and now has reached 50% survival.*

Optimism is not confined to the lay press. In 2003, then NCI Director Andrew von Eschenbach, announced that the NCI intended to "eliminate death and suffering" from cancer by 2015 (17). How can we hope to eliminate cancer deaths by 2015 when, 37 years after the U.S. war on cancer was launched, the cancer death rate in the United States has scarcely budged?

The reason that many people working in the cancer field believe that we are currently on track for conquering cancer by 2015 is that they are extrapolating progress based on survival data, not death data. It seems intuitively obvious that survival data and death data must be closely related, with deaths representing the residuum of cancer patients who are not survivors. This is simply not the case.

The purpose of this chapter is to explain how survival data create a false impression of medical progress. You might be wondering why it is important to show that we are losing our war against cancer. What is the value of a discouraging analysis? Our efforts to conquer cancer have focused on trying to enhance survival of patients with advanced common cancers (e.g., lung, colon, breast, prostate, and pancreas). These efforts have had only minimal value. If the public believes that cures for common, advanced cancers will soon be forthcoming, then alternative anticancer strategies will not be supported. The theme of this book, which is developed in the next several chapters, is that the only proven strategies for greatly reducing the number of deaths from the common cancers of adults are cancer prevention, precancer treatment, and early intervention. Of these three approaches, treatment of precancers is the least understood and least funded option.

2.2 Increased Cancer Survival without Increased Cancer Cure Rates

Laypersons think about cancer survival in terms of achieving a cure. **Oncologists** think about cancer survival in terms of the likelihood that a treatment will prolong the life of the patient. When oncologists think about increased survival, they are usually thinking about an increase in the length of time that a person lives with the cancer.

Treatments that increase cancer survival are tested in **clinical trials**. Today, the design of clinical trials has become standardized. A candidate treatment is tested in three phases:

Phase I trial—Trial designed to develop the safe dose and toxic effects of a candidate drug.

Phase II Trial—Clinical trial designed to determine if a drug has any efficacy (i.e., if it should proceed to a larger, Phase III trial), and, in the case of a cancer trial, which cancers are most likely to respond to the drug. In Phase II, additional information on the dosage and toxicity of the drug is collected.

Phase III Trial—Clinical trial designed to test a new treatment protocol, measured against the current standard treatment.

Because clinical trials measure the length of survival of cancer patients following treatment, trial results are usually reported in terms of increased survival. Here is an example. The NCI published a bulletin that featured an article on Gemcitabine, a chemotherapeutic agent for pancreatic cancer (18). The article quotes Dr. Helmut Oettle, who reports that "We have shown that this treatment more than doubles the overall survival five years after treatment." Does this mean that Gemcitabine doubles the odds that a person with pancreatic cancer will be cured of his/her diseases? No. For this study, only pancreatic cancer patients who qualified for surgical **resection** of their cancer were selected for adjuvant treatment with Gemcitabine. About 80% of pancreatic cancer patients do not qualify for surgery, because their cancer is widespread at the time of diagnosis. The patients who received Gemcitabine were all in a group of patients with a relatively favorable prognosis. In the Gemcitabine group, 21% of patients survived five years, compared with only 9% in the control group. Dr. Oettle did not exaggerate his findings. However, the jump in five-year survival may have been due to a small increase in length of survival of patients who had lived almost five years with their cancer (without Gemcitabine treatment). As it turns out, the median survival in the Gemcitabine group was 22.8 months compared with 20.2 months in the control group. Median survival, namely, the length of time during which half the patients in the study will die, is a better indicator of treatment effectiveness than the percentage of patients who survive five years. Gemcitabine extended the median life expectancy by 2.6 months in those patients who were eligible for surgical treatment and who survived the surgery long enough to receive six months of postsurgical treatment with Gemcitabine. The study did not address the question of whether Gemcitabine cured patients with pancreatic cancer. In addition, a Phase II trial is conducted to determine whether a treatment is likely to have value for patients. Phase II trials are too small to determine, with any confidence, the effectiveness of a treatment. That determination comes from a larger, longer Phase III trial.

Here is how a skeptic would summarize the results of the Gemcitabine trial: "In a preliminary analysis, announced by the same scientists who conducted the clinical trial, performed on a small number of patients in a subset of pancreatic cancer patients with a relatively favorable prognosis, median survival was extended 2.6 months. These results need to be reexamined in a larger, Phase III trial, and confirmed by posttrial clinical experience." Are the scientists who conducted the Gemcitabine study guilty of fraud? No. The scientists who conducted this trial, as well as those who conduct countless other clinical trials worldwide, are dedicated humanitarians. They are doing the best they can to make incremental advances against cancer. Their job is to report the merits of their studies. The healthcare community's job is then to interpret those findings with a degree of skepticism.

2.3 Effective Treatments That Do Not Cure

Laypersons often think of effective treatment as a drug that cures a disease. Oncologists think of effective treatment as an available option that has some clinically proven positive effect.

Bevacizumab (developed and sold by Genentech as Avastin) is one of the most popular cancer drugs in the world and has been heralded as a wonder drug. It can cost $50,000 to $100,000 per year of use. One of the most exciting features of Avastin is that it can potentially treat any kind of cancer. Avastin is an antibody that works by attaching to the vascular endothelial growth factor (VEGF), thus, reducing the ability of tumors to vascularize and grow. In responsive cancers, studies indicate that it extends life by up to four months (19).

You may not be impressed by a drug that can extend life by up to four months in some responsive patients. To the best of my knowledge, nobody has claimed that Avastin will cure advanced cancers. Why then is Avastin popular? A well-marketed drug that promises hope for cancer patients can have enormous appeal to desperate patients and their oncologists.

Are the results of clinical studies skewed in favor of the corporate sponsors of the trials? In a fascinating **meta-analysis**, Yank and coworkers wanted to know whether the results of clinical trials conducted with financial ties to a drug company were biased toward favorable results (20). They reviewed the literature on clinical trials for antihypertensive agents and found that ties to a drug company did not bias the results. The authors found, however, that financial ties to a drug company are associated with favorable conclusions. This suggests that regardless of the results of a trial, the conclusions published by the investigators were more likely to be favorable, if the trials were financed by a drug company. This should not be surprising. Two scientists can look at the same results and draw entirely different conclusions. How could investigators, financed by a drug company, not be influenced by their benefactors when they interpret their results?

2.4 Biases in Clinical Trials

When we interpret the results of clinical trials, we need to understand that biases are almost always introduced into the design, execution, and analysis of these trials. Most **trialists** do everything they can to reduce biases, but the "perfect" trial is seldom, if ever, achievable. Some of the biases that can diminish the value of clinical trials are in List 2.4.1. The biases in this list that are not covered in this chapter are described as individual entries in the Glossary.

The list of biases is long, but each type of bias has contributed to commonly held misconceptions about cancer treatment. How can we possibly choose a new direction for cancer research, if we cannot fully appreciate the ways that we have deceived ourselves over the past few decades?

List 2.4.1 Partial listing of biases in cancer survival data.

1. **Stage assignment bias**—diagnostic or screening tools that shift the proportion of people at different disease stages
2. Lead-time bias—extending time-after-diagnosis without changing date of death
3. **Population bias**—exclusion of important subpopulations
4. **Statistical method bias**—different methods in different studies of the same treatment
5. **Demographic bias**—different demographics over the time interval of the study
6. **Measurement bias**—inability to measure accurately clinical or biological study parameters
7. Record bias—medical records can be incomplete or otherwise flawed
8. **Reabstraction bias**—gaps in records often require reabstraction from other sources, and there may be biases in the way that additional information is collected for certain subpopulations of patients
9. **Comorbidity bias**—survival may be determined by processes other than the cancer under study
10. Diagnosis bias—when the patient has a benign condition that is erroneously diagnosed as cancer
11. **Second trial bias**—also called patient selection bias;second trials of the same drug or procedure commonly produce better survival results than the first trial, because the clinicians become more adept at selecting patients who will benefit from the treatment
12. **Marketing bias**—if there's money at stake, even minimal benefit can be spun into major advances; four-month extensions in life can be promoted as miracles
13. **Stage treatment bias**—finding an improvement that is effective for a small subset of people with a cancer can be falsely interpreted as a measure that enhances survival for everyone
14. **Apples-oranges bias**—sometimes improvements in survival can't be determined because the objectives of different studies may not be comparable
15. **Underreporting bias**—only reporting trials that demonstrate improved survival
16. **Confounder bias**—sometimes factors may improve health and survivability without being part of a treatment; for example, statins may increase survival by reducing the risk of heart disease in cancer patients without having any direct effect on the patient's cancer

2.5 Stage Treatment Bias

If you carefully select a **stage** of disease that is successfully treated by a particular treatment protocol, you can exaggerate the benefits of your treatment by ignoring disease stages for which your treatment is ineffective.

An example is the use of **prostatectomy** for **prostate cancer**, a procedure that is credited with a high cure rate. Prostatectomy is only performed on patients with tumors confined to the prostate. If the prostate cancer has metastasized to lymph nodes in the region of the prostate or to distant organs, then prostatectomy is contraindicated. Why? If the cancer has spread beyond the prostate, then removing the prostate will not benefit the patient. The cancer will simply grow from its metastatic deposits.

Prostate cancer confined to the prostate is often indolent. By age 80 to 90, 70% to 90% of men have prostate cancer confirmed at **autopsy** (21, 22). This indicates that prostate-confined cancer is a very common disease that kills only a small proportion of affected individuals. Because prostatectomy is only performed on men whose prostate cancer is believed to be confined to the prostate, the cure rate is high. Restricting treatment to patients who have a stage of disease that is indolent in most cases virtually guarantees high survival rates.

2.6 Stage Assignment Bias

Of the sources of bias listed, **stage assignment bias** is the most subtle and the most difficult to understand. It is also the most important bias introduced by the most modern cancer trials, which are often aimed at developing stage-specific treatments for groups of patients with common cancers.

Suppose every patient with type X cancer is staged into one of two groups (I and II). The stage I group has no evidence of distant **metastases** at the time of diagnosis and has a 40% chance of five-year survival under the standard treatment protocol. Patients are put into the stage II group if they have distant metastases at the time of diagnosis. Their chance of having a five-year survival under the standard treatment is 2%.

Here's the scenario. Professor Rads, at the University of Goodcare, has recently developed a very sensitive imaging device that can detect small metastases that would be undetectable by less sophisticated devices. In the next clinical trial for treatment of cancer X, Professor Rads tests each trial candidate with his new device. He finds that about half the patients who would otherwise be assigned to stage I (no metastases) have metastases with his sensitive machine. With this information, these erstwhile stage I patients are reassigned into stage II.

A clinical trial is conducted with the standard treatment. The stage I group is now much smaller than the stage II group. When the trial is complete, we find that the five-year survival for the stage I group is now 80% (up from 40%). The five-year survival for the stage II group is now 2% (the same as before). The newspaper headline following the trial is, "New Imaging Discovery Yields 100% Improvement in Survival for Stage I Cancer X."

Actually, the clinical trial, as described, yielded no improved survival for any patients. All it accomplished was to correctly reassign some of the stage I patients into the low-survival stage II group. The apparent improvement in survival in stage I cancer patients was the result of more accurate **staging** of patients in the low-risk category of disease.

What does this mean? Was the clinical trial a fraud? Did it accomplish nothing at all? No, accurate staging of patients with cancer is an absolutely crucial step in the development of new, effective anticancer regimens. Assessing the effects of a new chemotherapeutic agent on a group with heterogeneous disease is impossible. The beneficial effects of a new drug on stage I cancer patients might be lost in a study in which the stage I group is mixed with patients with stage II cancers that might not be responsive to the regimen.

By accurately staging patients with cancer, trialists can test and develop drugs that are most likely to benefit patients in a particular stage of disease. In fact, that is the current approach to cancer trials: subdividing cancer patients into well-defined clinical subsets and fine-tuning new anti-cancer regimens to maximize their effectiveness for each subset of disease. Progress in cancer staging, however, should not be confused with improved cancer survival.

2.7 Lead-Time Survival Bias

Suppose there were a cancer, cancer Y, that is uniformly deadly. Once it is diagnosed, the average survival, after the best available treatment, is three years. Nobody who has this cancer lives beyond five years.

Dr. Detecto is a pathologist who has invented a very sensitive method for detecting cancer Y at a very early stage. Dr. Detecto can detect cancer Y a full four years earlier than any previous method of detection. Unfortunately, there is no effective treatment for cancer Y, even when it is detected early. All patients with cancer Y will die. Because cancer Y patients are now detected four years earlier, the natural course of disease results in an expected death 7 years (3 years plus the 4 years lead time) later. When we study five-year survival after diagnosis, we find that the five-year survival is now 90%. The newspaper headline reads, "New, improved detection technique for cancer Y improves five-year survival from 0% to 90%."

Of course, detecting the cancer four years earlier only increased the time between diagnosis and death. It did not extend, by even a single minute, the age at death of patients with cancer Y. Has Dr. Detecto made a useless discovery, and are the survival data fraudulent? No. Tumors are best treated when they are detected early. In the case of cancer Y, there was no immediate benefit for early detection. Nonetheless, the set of early cancers provided cancer researchers with a group of tumors that might have an improved response to alternate cancer therapies. This would require clinical trials. The survival data from the early-diagnosed group can be very misleading, unless lead-time bias is eliminated.

2.8 Population Bias

Some populations **accrue** less easily into clinical trials than other groups. Many clinical trials exclude children and pregnant women. In such cases, drug effectiveness and safety cannot extend to these excluded groups. Third-party payers may refuse to cover the costs of drugs for pregnant women and children because of the lack of trial evidence indicating that the drugs are safe and effective in these groups.

Population bias effects every population that is not included in study populations (23). A good example comes from our interpretation of prostate specific antigen (PSA) values in men. PSA is the most important screening test for prostate cancer. PSA levels less than 4 ng/mL indicate low risk of prostate cancer. PSA levels greater than 4 ng/mL indicate high risk and prompt a diagnostic study. These clinically accepted ranges of PSA levels were adapted from data collected from a community-based Minnesota population consisting entirely of white men (24). In discussing the importance of age-specific ranges, the Mayo Clinic investigators acknowledged that African-Americans and Asians were not included in their study (25).

Do the screening criteria, developed for PSA ranges in a white population, fit the normal ranges for African-Americans and Asians? In a review of PSA levels among white men and black men conducted by Sawyer and coworkers, studying 30,000 PSA values collected for about 3,000 subjects, African-Americans had a higher range of PSA levels than white Americans (26). A separate study of PSA levels in Asian men found a lower range of PSA levels in this population (27). A population bias occurs when PSA test results on African and Asian men are interpreted using criteria established for a population of Caucasian men.

2.9 Diagnosis Bias

Diagnosis bias occurs when a trial group contains patients who do not all have the same diagnosis. This bias distorts survival outcome after treatment. For example, if patients were accrued into a breast cancer study based on erroneous diagnoses rendered by an incompetent pathologist, then the trial group might contain patients who have **benign** breast disease. This would produce an inflated survival rate, compared to earlier studies that did not mistakenly include healthy patients.

Most modern clinical trials impose strict pathology reviews in an effort to reduce the effect of diagnosis bias.

More commonly, treatment groups are tainted when a new subset of patients is added to the original group through enhanced detection and diagnosis of a biologically distinctive variant of disease. For example, there has been a many-fold increase in the detection and diagnosis of **ductal carcinoma *in situ*** (DCIS) of the breast over the past few decades, thanks to mass mammographic screenings (28). DCIS has a very good prognosis. Only a small percentage of patients with DCIS progress to invasive ductal carcinoma.

As the proportion of breast cancer patients with diagnosed DCIS increases, the survival of the group that includes DCIS cases also increases. By increasing the number of patients with good-prognosis breast cancer, improved survival can be achieved without an improvement in therapeutic protocol.

The remedy for diagnosis bias is to subdivide your study groups carefully. This remedy has its own problems. If the first study did not carefully separate patients by their prognostic subtypes, how can you compare prior results with future results (in which the subtypes were separated). Also, if you break up groups by their tumor subtypes, does each group contain a sufficient number of patients to produce statistically meaningful results?

2.10 Living with Statistical Ambiguity

John P.A. Ioannidis is the chair of the Clinical and Molecular Epidemiology Unit at the University of Ioannina School of Medicine and Biomedical Research Institute in Greece. In a provocative article entitled "Why most published research findings are false," he points to some common misinterpretations that pose as clinical facts (29). These include post hoc subgroup selection and analyses (i.e., **cherry-picking** a subgroup that qualifies for statistical significance); changing clinical group inclusion or exclusion criteria and disease definitions after the trial has concluded; selective or purposely distorted reporting of results; data dredging (sifting through study data, searching for outlier groups); and for multicenter studies, reporting the significant findings from some centers and ignoring negative results from other centers (29, 30). Dr. Ioannidis has published a list of conditions that reduce the likelihood that a research finding is valid (List 2.10.1).

Despite their flaws, clinical trials are vital for medical progress. Those interested in understanding cancer need to develop an attitude that tempers their desire to find new, effective cures, with an understanding that clinical trials are simply another type of experiment. Like all experiments, they can be poorly designed, misinterpreted, invalid for under-represented patient subpopulations, unrepeatable, and falsified. The best validation of predictive tests comes from continuous clinical correlations with patient outcomes in medical centers where many different types of patients (male, female, different nationalities, different ages, concurrent diseases, multiple medications) are managed. All clinical trials must be validated through post-trial clinical outcome data.

Human nature tends to celebrate success and bury failure. When a clinical trial produces negative results (fails to show improved survival), there may be little enthusiasm to publish the work. Sponsors of negative studies may be disinclined to rally the cancer research community and the public around their negative results. Dickerson and Rennie have written, "The fact that some trial results are never published would not be a problem, except that there is good evidence that the results from unpublished trials are systematically different from those of published trials" (31). NIH provides a free website for federally funded and private organizations to register their clinical trials (32). The data from all clinical trials must be made publicly available, so that every conclusion can be discovered, scrutinized, and debated.

List 2.10.1 When to doubt a published research finding (29).

1. Small studies are less likely to produce true research findings than large studies.
2. Small effects are less likely to be true than larger effects.
3. Research findings are more likely to be true in confirmatory studies (such as phase III trials that confirm observations made in phase I and phase II trials) than in hypothesis-generating studies.
4. The greater the flexibility in design, definition, and measured outcome, the less likely that the research findings will be true.
5. The greater the financial and other interests in a study, the less likely that the results will be true.
6. The hotter the scientific field, the less likely that the results will be true.

2.11 Time Is Not on Our Side

Modern clinical trials are long and expensive. The process of testing a prospective new drug can take many years. In the realm of cancer trials, the Prostate, Lung, Colorectal, and Ovarian Cancer Screening Trial (PLCO, NIH/NCI trial NO1 CN25512) serves as an example. PLCO is a randomized, controlled cancer trial. Between 1992, when the trial opened, and 2001, when enrollment ended, 155,000 women and men between the ages of 55 and 74 joined PLCO. Screening of participants and the collection of follow-up data will end around 2016. The purpose of the study is to determine whether screening reduces mortality from these cancers (33). We need 24 years to answer this question.

Although prospective trials are often considered the only way of determining the efficacy and safety of new treatments and diagnostic tests, the public may legitimately ask whether society has the time, money, and patience for these studies. Those engaged in clinical trials may well ask themselves whether some clinical trials should be replaced by new, innovative models, producing clinically sound results in less time and for less money.

There are literally thousands of different neoplasms in humans, all requiring stage-stratified clinical trials to find effective treatments. Can society test all the potential chemotherapeutic agents (most of which will turn out to be ineffective) on all these cancers?

Chapter 2 Summary

When oncologists discuss improved outcomes for advanced-stage common cancers, they are seldom referring to a patient's chances of achieving a full recovery. More often than not, they are discussing the length of time that a patient is expected to live following the diagnosis of the cancer. Although survival times are incrementally increasing, clinical trial data tell us nothing about improvements in cancer death rates.

Clinical trials are designed to provide information that applies to a selected group of patients, in a specific stage of disease, with specific and predetermined endpoint measurements. You cannot generalize about populations using data taken from a clinical trial. When you do, you inevitably encounter assorted biases that distort your conclusions. When assessing any clinical trial, there are a number of questions that you should always ask (List 2.11.1).

List 2.11.1 Minimal information needed to assess drug efficacy announcements.

1. What exactly is the trial population? What fraction of the total population of patients with cancer would be qualified for treatment?
2. Is the treatment intended to cure the patient?
3. Is the trial a phase I, II, or III trial?
4. Is the trial registered with http://www.clinicaltrials.gov (32)?
5. Have the trial researchers disclosed any and all conflicts of interest (such as funding by a drug company or financial rewards resulting from a positive outcome)?
6. Do the trialists obey a strict policy for reporting trial results, whether results are positive or negative (31)?
7. Are trial raw data disclosed to the public?
8. What are the mean and median survival times for the treated group and the control group?
9. Has the effectiveness of the drug been confirmed in posttrial outcome analyses on a large population collected from many hospitals?

CHAPTER 3

Why We Have Not Learned How to Cure Advanced Common Cancers

Medicine can only cure curable diseases, and then not always.

—Chinese proverb

3.1 Background

The general approach to funding cancer research is the same now as it was in the early 1970s; attack cancer tumor by tumor, dividing available funds in rough proportion to the number of people who die from each tumor (List 3.1.1).

The top five cancer killers (lung, colon, breast, pancreas, and prostate) account for 57% of all cancer deaths. That the cancers that kill the most people are the same cancers that receive the bulk of funding from the U.S. National Cancer Institute is no surprise (List 3.1.2). Funding for the top 15 cancer killers receive U.S. funding roughly in proportion to the number of Americans they kill. Other cancers receive much less funding. For example, stomach cancer receives $12 million and uterine cancer receives $16.6 million. **Rare tumors** receive a tiny fraction of the cancer research budget.

The only problem with this straightforward approach is that it has failed. Despite decades of funding, we still do not know how to cure the most common advanced cancers occurring in humans. New discoveries in cancer genetics have highlighted the large number of **mutations** and genetic alterations that accumulate in the common cancers of humans. These advances in knowledge have not led to cures for advanced cancers or to large increases in the length of time that people live after diagnosis of an advanced cancer. Moreover, because the advanced common cancers are genetically complex, there is little hope of finding a miracle drug that can cure these cancers by targeting a single molecule or a specific molecular pathway.

List 3.1.1 Age-adjusted U.S. death rates and trends for the top 15 cancer sites, expressed as deaths per 100,000 population (34).

All sites	189.8	Liver and intrahepatic bile duct	5.0
Lung and bronchus	54.1	Ovary	5.0
Colorectal	18.8	Esophagus	4.4
Breast	14.1	Brain and other nervous system	4.4
Pancreas	10.6	Urinary Bladder	4.3
Prostate	10.1	Kidney and Renal Pelvis	4.2
Leukemia	7.4	Stomach	4.1
Non-Hodgkin lymphoma	7.3	Myeloma	3.7

List 3.1.2 Cancer funding by NCI, by cancer site, expressed as millions of dollars spent in 2007; data from Office of Budget and Finance (35).

Lung cancer	$226.9 million
Colorectal cancer	$258.4 million
Breast cancer	$572.4 million
Pancreatic cancer	$73.3 million
Prostate cancer	$296.1 million
Leukemia	$205.5 million
Non-Hodgkin lymphoma	$113.0 million

When you think about the real successes in treating advanced, metastatic cancers, virtually all success has occurred in the realm of rare tumors: pediatric **neoplasms**, certain **lymphomas** and **leukemias**, germ cell tumors, and gastrointestinal stromal tumors (GISTs).

The purpose of this chapter is to demonstrate that there are fundamental biological differences between rare tumors and common tumors. These biological differences account for our failure to cure common cancers and our success at curing several rare cancers. In later chapters, we will see that the same properties that account for the biological differences between rare and common tumors also account for the biological differences between precancers and the cancers into which they develop.

3.2 What Causes Common Cancers?

The human body can be envisioned as a topological donut (Figure 3.1). Like donuts, humans have a continuous surface. We can think of our skin as the outer-edge surface of the donut. We can think of our alimentary tract as the inner-edge surface of the donut that lines the donut hole. Our outer-edge surface is lined by epithelial squamous cells of skin epidermis. Our inner-edge surface is lined by epithelial **enterocytes** of the gastrointestinal **mucosa**.

Inside the donut is the pastry, corresponding to the nonsurface cells of the human body. The nonsurface cells of the human body are composed largely of **connective tissues**. The connective tissues include fibrous tissue, adipose tissue, muscle, bone, and vessels. These tissues derive from an embryonic layer called **mesoderm**, sandwiched between the two embryonic layers that produce the surface of the human body.

Virtually all exposure to toxic and carcinogenic chemicals takes place on the donut surfaces (skin and gastrointestinal tract) and the epithelial organs that bud off these surfaces (lungs, pancreas, liver, breast, prostate). Because tissue surfaces contain the cells that are most exposed to **carcinogens**, it is no surprise that most human cancers are tumors of surface cells (not internal cells). Just two cancers arising from the outer surface cells (squamous cell carcinoma and basal cell carcinoma) account for over one million new tumors each year in the U.S., roughly equal to the number of all the other tumors of the body combined. Tumors of tissue surfaces account for over 95% of the tumors that occur in humans.

Cells lining the surface of the body are constantly dividing. Presumably, the reason that surface cells divide more than the internal cells is that they need to replace

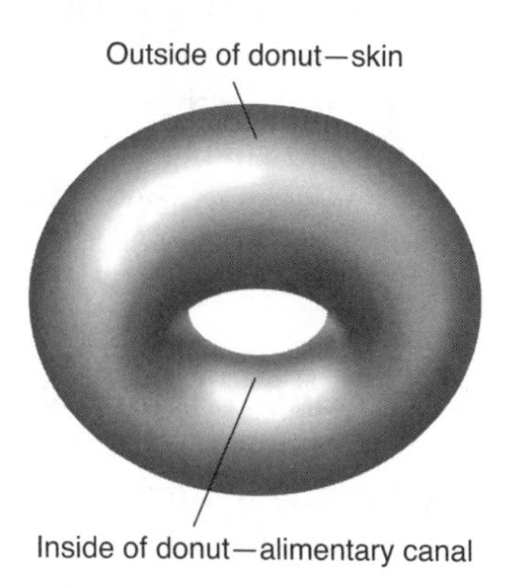

FIGURE 3.1 Donut version of human body.

List 3.2.1 Features of common cancers.

They have many different causes.

They are genetically very complex (i.e., have many small and large genetic aberrations, including *aneuploidy*, an abnormal number of chromosomes). Colon cancers contain about 11,000 **genomic alterations** (36).

They are heterogeneous (i.e., many different cancer subclones in the same cancer).

Death rates from the most common advanced cancers have been virtually unchanged since the war on cancer was declared in 1971.

themselves after exposure to toxins. The inner cells are protected from exposure to toxins, so their rates of cell death and cell replacement are low. Only dividing cells are targets for the early mutational steps of carcinogenesis, because dividing cells can pass unrepaired mutations to their progeny. Cells that divide slowly can repair DNA mutations produced by carcinogens. Cells that never divide have no progeny and are not targets of carcinogens. This is one more reason why rapidly dividing surface cells of the human body are much more likely to give rise to tumors than internal cells of the body.

For these reasons, most cancers are epithelial tumors that derive from surface cells of the body. These cells are constantly accumulating genetic alterations from continued exposure to carcinogens in the environment. Eventually, one or more of the mutations is likely to adversely affect the ability of cells to repair DNA. This results in **genetic instability** and greatly accelerates the accumulation of additional mutations.

Common tumors that develop from surface epithelial cells are caused by the many different carcinogens in our environment, and they are characterized by many different genetic lesions within each tumor and among different tumors of the same type. We can summarize these realities of common cancers (List 3.2.1).

3.3 Age-Based Successes In Cancer Treatment

Childhood cancer is rare, and every cancer occurring in a child is a rare cancer. The pediatric age group has had the largest drop in cancer death rate since 1950 (List 3.3.1). The overall cancer death rate (all ages) is 5.8% lower in 2005 than in 1950. Age groups above 65, for which cancer is much more common, have had an increase in the cancer death rates.

List 3.3.1 Cancer death rates stratified by age, all races, males and females, included (8). Rates are per 100,000 population and are age-adjusted to the 2000 U.S. population.

Age Group	1950	1978	2005
0–4	11.1	4.6	2.2
5–14	6.7	4.1	2.5
15–24	8.6	6.1	4.1
25–34	20.4	14.2	9.1
35–44	63.6	50.7	32.8
45–54	174.2	179.6	118.3
55–64	391.3	428.9	329.7
65–74	710.0	803.4	748.8
75–84	1167.2	1204.1	1265.1
85 and over	1450.7	1535.3	1643.7
All Ages	195.4	204.4	184.0

3.4 Why We Have Succeeded in Curing Some Rare Tumors

Cancer deaths among adults are about 500 times greater than childhood cancer deaths, and the high rate of deaths in adults has barely budged in over half a century. Biological differences between rare and common tumors account for the successes in treating some rare tumors of childhood.

Rare tumors tend to arise from the inside cells of the body, not from surface cells. Rare tumors include many tumors of infancy and childhood, and almost none of them develop from surface cells. Pediatric tumors develop quickly, unlike tumors occurring in adults, which typically develop many years after exposure to a carcinogen.

Congenital neoplasms (rare tumors present at birth) have, at most, nine months to develop. The long development phase of common tumors contributes to the accumulation of numerous genetic alterations that are present in all common cancers (and present in low numbers in **congenital neoplasms**).

There are few causes for any particular rare tumor (that is why a rare tumor is rare). The known cause for one rare tumor is likely to be the cause for all cases of that rare tumor. The cause might be an inherited mutation, as is the case for inherited **retinoblastoma**. The cause might be a single exposure to an identified carcinogen at a documented moment in time, as in gestational exposure to diethylstilbestrol resulting in clear cell **adenocarcinoma** of the cervix in adolescent girls.

List 3.4.1 Tumors that can be cured with chemotherapy (37)

Choriocarcinoma
Acute lymphocytic leukemia of childhood
Burkitt lymphoma
Hodgkin lymphoma
Acute promyelocytic leukemia
Large follicular center cell (diffuse histiocytic) lymphoma
Embryonal carcinoma of testis
Hairy cell leukemia (probable)
Seminoma

Rare tumors are likely to have a single cause, a single aberrant pathway, or a single inherited gene. It is possible to arrest the growth of a rare tumor by targeting a single identified genetic alteration. Knocking out one growth pathway in a heterogeneous population of genetically diverse cells, such as we find in advanced common tumors, will not stop the growth of the cancer.

We can cure a few of the rare tumors (List 3.4.1).

The new anticancer drugs, targeted at specific pathways activated in tumors, are most effective against rare tumors, characterized by simple, well-characterized mutations and **translocations**. Developing a cure for common cancers before we have developed a variety of approaches to curing genetically simple cancers is unlikely. Sometimes you need to crawl before you can walk.

3.5 What the Rare Tumors Tell Us about the Common Tumors

We spend our cancer research money on the same diseases that we spend our cancer care dollars on: lung cancer, colon cancer, breast cancer, and prostate cancer. But if you look for tumors that have led to breakthroughs in our understanding of cancer biology, you find a very different list: gastrointestinal stromal tumor, **chronic myelogenous leukemia**, **acute lymphocytic leukemia**, **Hodgkin lymphoma**, and **seminoma**.

Every rare tumor (and there are many) reveals new secrets (List 3.5.1).

Many of the greatest advances in our understanding of cancer have come through the study of familial cancer syndromes. Although familial cancers are almost always rare, they comprise a large group of syndromes, a few of which are in List 3.5.2.

List 3.5.1 Secrets of rare tumors.

Why is the tumor rare?
What special conditions must prevail before the rare tumor will occur?
How does the ordinary phenomenon of common tumor development compare with the extraordinary phenomenon that causes rare tumors to develop?

List 3.5.2 Familial syndromes associated with neoplastic development.

101000 Acoustic schwannomas, bilateral
113705 BRCA1 breast cancer, type 1 included
114400 Lynch cancer family syndrome II
115310 Carotid body tumors and multiple extraadrenal pheochromocytomas
120435 Colorectal cancer hereditary nonpolyposis type 1
130650 Wiedemann-Beckwith syndrome
131100 Multiple endocrine neoplasia type 1
132700 Cylindromatosis, familial
133510 **Xeroderma pigmentosum** complementation group B
135150 Birt-Hogg-Dube syndrome, fibrofolliculomas with trichodiscomas and acrochordons
151623 Li-Fraumeni syndrome 1
155240 Thyroid carcinoma, familial medullary
155600 Dysplastic nevus syndrome, hereditary B-K **mole** syndrome
158320 Muir-Torre syndrome, cutaneous sebaceous neoplasms and multiple keratoacanthomas with gastrointestinal and other carcinomas
158350 Cowden syndrome, multiple hamartoma syndrome, Lhermitte-Duclos disease included
160980 Carney myxoma-endocrine complex, Carney syndrome
162200 Neurofibromatosis type 1
175200 Peutz-Jeghers syndrome
188550 Familial nonmedullary thyroid cancer
191100 Tuberous sclerosis
193300 Von Hippel-Lindau syndrome
208900 Ataxia-telangiectasia
210900 Bloom syndrome
227650 Fanconi anemia
235200 Hemochromatosis
276300 Turcot syndrome, malignant tumors of central nervous system associated with familial polyposis of colon
278700 Xeroderma pigmentosum I
308240 Lymphoproliferative disease, X-linked
308940 Leiomyomatosis, esophageal and vulvar with nephropathy
600185 BRCA2 breast cancer, type 2 included
600376 Osler-Rendu-Weber syndrome 2
601650 Glomus tumors, familial 2
604287 Carney triad
605365 Breast cancer, type 3
605462 Basal cell carcinoma, multiple
605839 Leiomyomatosis and renal cell cancer, hereditary
606690 Lymphangioleiomyomatosis
606719 Familial atypical multiple mole melanoma-pancreatic carcinoma syndrome

Each inherited neoplasm syndrome is accompanied by an OMIM (Online Mendelian Inheritance in Man) number (38). The OMIM number can be used to retrieve the updated OMIM record for the syndrome (39). Additional information for people with rare diseases is available from the National Organization for Rare Diseases (40).

One must appreciate that there are hundreds of inherited cancer syndromes, while there are just a few dozen common cancers. When we examine the long list of inherited cancers, we can draw several important conclusions (List 3.5.3).

In most cases, the same genetic **lesion** that occurs in the **germline** of rare tumor syndromes is also present in some cases of sporadically occurring (nonfamilial) common cancers (List 3.5.4). In tumor after tumor, the genetic lesion present in sporadically occurring cancers would not have been found without prior knowledge of the syndromic gene (the gene responsible for the rare inherited condition).

In the 1930s through 1950s, thorotrast, a colloidal suspension of radioactive thorium dioxide, was commonly used as a radiological contrast medium (46). This medium is absorbed into the Kupffer cells of the liver. Kupffer cells are specialized phagocytic endothelial cells that line the hepatic sinusoids. Once absorbed into Kupffer cells, the radioactive compound emits alpha particles. Emission continues at an exponentially

List 3.5.3 General observations on the long list of inherited neoplastic syndromes.

1. In many cases, neoplastic changes occur in only a small minority of the patients who have inherited a cancer gene.
2. Inherited cancer genes are compatible with normal fetal growth and development. The cancer gene is developmentally silent, with the exception of specific abnormalities and predispositions characteristic of the syndrome. This means that cells can sometimes contain cancer genes without becoming cancerous.
3. Although persons with inherited cancer syndromes develop normally, with the exception of specific syndrome-associated abnormalities, those exceptional abnormalities indicate that cancer genes play narrow roles in development.
4. Many inherited cancer syndromes are associated with commonly occurring tumors. The Li-Fraumeni syndrome is associated with breast cancer, lung cancer, colon cancer, pancreatic cancer, and prostate cancer, as well as many rare tumors. Because not all people with common cancers have an inherited condition that caused the cancer, we can infer that common cancers must have multiple different causes.
5. Many cancer genes cause neoplasms that are specific for one germ layer (e.g., ectoderm, endoderm, mesoderm, or neural crest).
6. A single inherited cancer syndrome can cause all categories of neoplasms: cancers, benign tumors, hamartomas, and precancers. This implies that the cause of all these different types of neoplasms must share one or more steps in common.

List 3.5.4 A few examples of rare germline mutations that also occur in sporadic tumors.

1. Germline mutations of the p53 tumor suppressor gene are present in the rare Li-Fraumeni syndrome. A somatic p53 mutation is present in about half of all human cancers (41).
2. Families with germline mutations of the *KIT* gene develop gastrointestinal stromal tumors (GISTs). Somatic mutations of *KIT* occur in the majority of sporadic GIST tumors.
3. Germline *RET* **gene mutations** occur in familial medullary carcinoma of thyroid, and in sporadic cases of medullary carcinoma of thyroid (42), (43).
4. Germline *RB1* gene mutations occur in familial retinoblastoma syndrome and in sporadic cases of retinoblastoma (44).
5. Germline patched (*ptc*) gene mutations occur in basal cell nevus syndrome and in sporadically occurring basal cell carcinomas (45).
6. Germline *PTEN* mutations occur in Cowden syndrome and Bannayan-Riley-Ruvalcaba syndrome, two inherited disorders associated with a high rate of endometrial carcinomas. *PTEN* mutations are found in 93% of sporadically occurring endometrial carcinomas (38).

decreasing rate for the lifetime of the patient. Alpha particles have a short penetration distance, and the tissue cells most likely to be damaged are cells containing thorotrast and cells immediately adjacent to these cells. Tumors caused by exposure to radiological thorotrast are hepatocellular carcinoma, cholangiocarcinoma, and hepatic angiosarcoma. Hepatocellular carcinoma is a common tumor, but cholangiocarcinoma is rare, and angiosarcoma of the liver is extremely rare. Hepatic angiosarcoma was the sentinel tumor that warned workers in the fledgling field of radiology that thorotrast was causing cancers. Today, barium is used as a contrast agent, and the incidence of liver angiosarcomas has plummeted.

In 1974, representatives of the rubber and tire manufacturing industries reported that several angiosarcomas had occurred in three workers who worked at the site of vinyl chloride polymerization reactor vessels (47, 48). This news arrived several years after studies on rats had shown that aerosolized polyvinyl chloride was carcinogenic. Once again, the occurrence of just a few rare tumors, in the setting of a specific type of occupational exposure, was sufficient to alert the public that another carcinogen had entered the workpace environment (47).

In a landmark paper published in 1971 by Herbst and coworkers, the authors found an increase in the number of young women who developed an extremely rare cancer: clear cell adenocarcinoma of the cervix or vagina (49). The mothers of most of these young women had ingested a nonsteroidal synthetic estrogen (diethylstilbestrol, DES) during their pregnancies. Somehow, in utero exposure to a drug caused a specific rare tumor to occur in the offspring. The oldest patient was 27 years old at the time of diagnosis. The youngest was 7 years old. Ninety-one percent were 14 years of age or older

(50). Men exposed in utero to DES seemed to suffer no increased cancer risk (51). The findings of Herbst and coworkers, observing a rare tumor entirely by chance, were a major impetus for advancement in the field of perinatal **carcinogenesis**.

A rare tumor encountered today may be a common tumor in the future. Before smoking became a national pastime, bronchogenic **carcinoma** was a rare tumor. If the link between cigarettes and cancer had been established when the number of new tumors (and new smokers) was still small, a vigorous effort to curtail smoking might have staved off the current worldwide epidemic of lung cancer.

Most cancers that we can prevent, by simply removing a chemical from our environment, are rare tumors.

Why have we done so much better for rare tumors than for common tumors? Rare tumors are much more likely to have a single cause, a single carcinogenic pathway, a single inherited gene, or a single acquired **marker**, than any of the common tumors. When we look at past breakthroughs in cancer epidemiology, most are examples in which a small population was exposed to a specific carcinogen that produced a rare tumor (List 3.5.5).

Common tumors develop within a large, diverse population, not a population with a unique **genotype**. When the **prevalence** of a tumor is high, finding a single carcinogen to which all and only those cancer patients were exposed is difficult or impossible. When we study common cancers, we find multiple classes of genetic markers that characterize different examples of the same type of tumor. For example, colon cancers may contain any of several different classes of tumor suppressor genes resulting in loss of genetic stability (52).

If a single gene were responsible for most or all of the cases of a fatal disease, natural selection would have reduced the occurrence of the gene in the population. In almost every commonly occurring disease with a genetic component, we find different rare gene variants accounting for small subsets of the disease population (53). This has been

List 3.5.5 Epidemiologic breakthroughs in cancer research.

Hepatic angiosarcoma (thorotrast exposure in the 1940s, polyvinyl chloride exposure in the 1970s)
Mesothelioma (in World War II asbestos-exposed U.S. Navy shipbuilders, 20+ years after exposure)
Scrotal cancer (in British chimney sweeps in the eighteenth century)
Esophageal adenocarcinoma (late twentieth century, particularly among obese, middle-aged whites)
Oral cancer in teenagers (tobacco chewers, late twentieth century)
AIDS-related cancers (beginning about 1980)
Lung cancer (yes, this was a rare tumor in the nineteenth century)
Leukemia (from benzene exposure)
Thyroid cancer in Chernobyl (pulse exposure to a carcinogen)

true for Alzheimer's disease, Type 2 diabetes, and breast cancer. There is little reason to assume that common tumors have a common genetic cause, and there is little expectation that we can find all of the many gene variants that contribute to common cancers.

3.6 Connection between Rare Tumors and Precancers

In many cases, rare cancers express the minimal, essential genetic features of cancer. For purposes of finding a cancer cure, the ideal tumor is characterized by a common set of properties (List 3.6.1).

A good example of an (almost) ideal cancer is gastrointestinal stromal tumor (GIST). GIST is a very rare tumor that arises from the interstitial cells of Ramon y Cajal in the intestine. The interstitial cells of Ramon y Cajal aid in normal gut peristalsis. GISTs are characterized by a mutation in the *c-KIT* gene, which codes for a tyrosine kinase receptor. GIST tumors can be identified with an immunostain for CD-117, the c-KIT receptor kinase protein, which is strikingly elevated in GIST cells. GIST can be treated with Gleevec, a drug that targets the c-KIT tyrosine kinase receptor protein, disrupting the cellular pathway that drives the growth of GIST cells (54, 55).

Gleevec was the first **nontoxic**, **molecular-targeted** chemotherapeutic agent found to be effective in the treatment of a solid tumor (it had already been found to be useful in the treatment of chronic myelocytic leukemia) (56). Before Gleevec, there was no effective treatment for metastatic GIST tumors. Today, thousands of patients with GIST have benefited from this novel, modern treatment. This advance could not have been achieved if pathologists had not recognized that the GIST variant of gastrointestinal spindle cell tumors is a specific neoplasm, growing from a specific **cell type**, and having a narrow range of molecular alterations that drives its growth.

Life is never ideal. There are some cases of GIST characterized by a mutation that produces an altered protein that is not responsive to Gleevec treatment. There are some cases of GIST that develop resistance to Gleevec over time. Still, GIST shows us that when a tumor is characterized by a small number of genetic alterations, developing a cure by selectively modifying the cellular pathways that drive the

List 3.6.1 Properties of the ideal rare cancer.

Single genetic cause or single environmental cause
Single gene marker
Single pathway
Single chemotherapeutic target
Single cure that is effective on all cancers of the same type

malignant **phenotype** may be possible. When a tumor is complex, with hundreds or thousands of genetic changes, and with multiple tumor **subclones** each with different sets of genetic aberrations, the task of treating the tumor becomes very, very difficult.

Does this mean that our goal of curing all cancers is unachievable? Will we only be able to cure the very rare tumors that are characterized by a simple set of genetic changes?

We need to remember that every complex cancer began as a simple cancer. In Part 2 of this book, we will see that many precancers, unlike the cancers into which they develop, are often genetically simple lesions. This is particularly true for myeloprolferative and lymphoproliferative disorders and for the very earliest forms of epithelial precancers. In Part 3, we will see that lessons learned from rare tumors can be used to develop effective treatments for precancers.

Chapter 3 Summary

Although the bulk of cancer research funds are spent on cures for common cancers (lung, colon, breast, and prostate cancers), we will unlikely find cures for any of these cancers, in their advanced stages, anytime soon. We now know that these cancers are genetically complex. At this time, gene-targeted cancer therapies have limited use against tumors that harbor thousands of mutations. Though we have spent decades seeking cures for the advanced, common cancers, we have little to show for the effort.

The cancers that we have learned to cure are rare tumors, characterized by simple genetic abnormalities. Our best hope of finding cures for the commonly occurring cancers will almost certainly come from experience gained by curing rare tumors.

In later chapters, we will see that precancers have properties in common with rare tumors and these shared properties provide simple strategies for curing the precancers and for preventing precancers from developing into advanced cancers.

PART

Precancer Pathology and Biology

CHAPTER 4

Guide to Pathologic Examination of Precancers (for Laypersons)

The beginnings of all things are small.

—Cicero

4.1 Background

When I speak with biochemists and molecular biologists who work in cancer research, I find that many of them regard precancers as an abstract concept: the collection of unknowable cellular conditions that must occur before a malignant cell mysteriously comes into existence and spawns a new cancer. This is simply not the case. Precancers are small but observable lesions that can be biopsied and precisely diagnosed by pathologists. Cancers arise from precancers; they do not arise from normal cells that have suddenly become "malignant" through an unfortunate gene mutation.

The purpose of this chapter is to reveal the secrets of precancers that are known to every pathologist. You will learn how tissues are prepared for microscopic examination and how pathologists evaluate the histologic features of precancers and cancers. After reading this chapter, you will fully understand the information and arguments presented in the later chapters of this book.

4.2 Role of the Anatomic Pathologist

Pathologists, the people who render diagnoses by examining samples of diseased tissue, live in a secret world. They employ techniques and instruments that few people would use in the course of everyday life. They see things that nobody else, including other physicians, ever notice. They know the names of every different type of tumor, and they can identify those tumors in an instant. In 1858, Rudolph Virchow published

Cellular Pathology, describing the principles of pathology as they were known to Virchow and his contemporaries (57). These same principles have been studied for the past 150 years. Simply put, diseases result from changes that occur in our cells. By examining diseased cells under a microscope, we can identify the disease, and we can often find the cause and cure of the disease. Since the mid-nineteenth century, virtually every advance in medicine has required the careful pathologic examination of diseased cells. Cancers are diagnosed by biopsy, followed by microscopic examination. A biopsy is a sampling of tissue (Greek: *bios* = life; *opsis* = appearance).

All tissues removed from patients (e.g., by surgeons during operations, by dermatologists who sample small skin lesions, by **phlebotomists**, by patients themselves when they collect sputum or urine specimens) go to the pathology department, where they are examined. In many cases, samples of the tissues are fixed in **formalin**, then processed to produce paraffin-infiltrated tissue encased in a block of wax. These blocks, or cassettes (because plastic cassettes hold the paraffin block), are sliced into thin tissue sections and mounted onto **glass slides**.

The tissue slice is stained with chemical dyes that colorize the components of individual cells. The stained cells are examined under a microscope by an **anatomic pathologist**, a physician who specializes in evaluating cells and tissues under a microscope. Anatomic pathologists reach a diagnosis based on the correlation of clinical, gross anatomic, and microscopic features of the lesions.

Unlike radiologists, who look at visual representations of physical lesions, pathologists look at the actual cells taken from the patient. Provided that the tissues were properly fixed after removal from the patient, archived paraffin-embedded tissues, sometimes over 100 years old, are perfectly suitable material for modern research studies. About 25 million surgical pathology specimens are collected each year in the United States. Over the past century, a distributed archive of human tissue blocks has accumulated that has enormous value for biomedical researchers. In most cases, an anatomic pathologist can examine a stained sample of tumor and determine its precise diagnosis in a few seconds (comparable to the time that you would need to identify the *Mona Lisa* or to distinguish an apple from a banana).

4.3 Morphology of Malignant Cells

Tumor cells are different from normal cells because they have genetic alterations. Because genomic DNA is confined to the **nucleus**, you might expect the nucleus of a malignant cell to have a different morphology than the nucleus of a normal cell. This is in fact the case. Observations made on cancer cell nuclei form the basis of diagnostic medical cytology (the study of cells).

Cytologists practice their profession by observing collections of stained cells obtained from body fluids (such as **pleural** or peritoneal **fluids** or urine), from scrapings of epithelial surfaces (such as **Papanicolaou [Pap] smears** from cervix), or from fine-needle aspirations of internal or external nodules. A representative cell population of just a few cells from a **neoplasm** can suffice for a cytologist to render a diagnosis.

Malignant cells have a different morphology than their normal tissue counterparts (the normal cells that look most like the malignant cells, and from which the malignant cells probably arose). Let us examine a fine-needle aspirate of lung. **Fine-needle aspirates** are cytologic preparations made from cells removed from inside a tissue.

Normal bronchial **epithelial cells** have a uniform round or oval nucleus at one end of the cell (base). The other end of the cell (apex) has a flat border from which many short **cilia** extend, like eyelashes from an eyelid (Figure 4.1). When an epithelial cell lies flat on a glass slide, you can see the full length of the cell, with the cilia at one end and the nucleus at the other. When cells lie on the glass slide *en face* (with the full length of the cell perpendicular to the plane of the slide), we see polyhedral cells, each containing a central nucleus.

The clumped normal cells are geometrically spaced, much like the cells in a honeycomb. Compare this with the morphologic features of a clump of malignant cells (Figure 4.2).

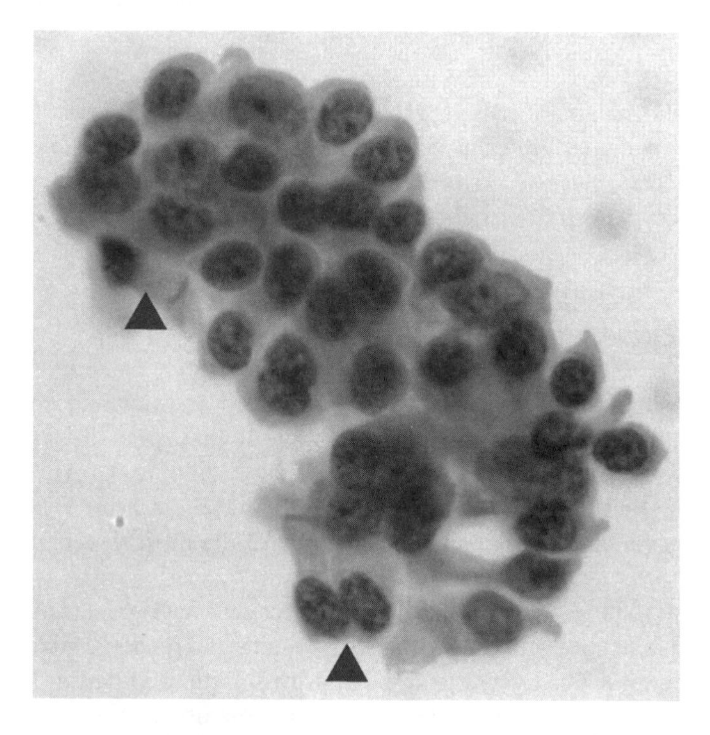

FIGURE 4.1 Normal bronchial epithelial cells. The upper left triangle points to an epithelial cell with nucleus on the left and a tuft of thread-like cilia pointing right and downward. The lower-middle triangle points to two epithelial cells with nuclei below and cilia pointing upward. Most of the cells in the clump are viewed *en face*, with oval nuclei and cilia lying out of the plane of focus.

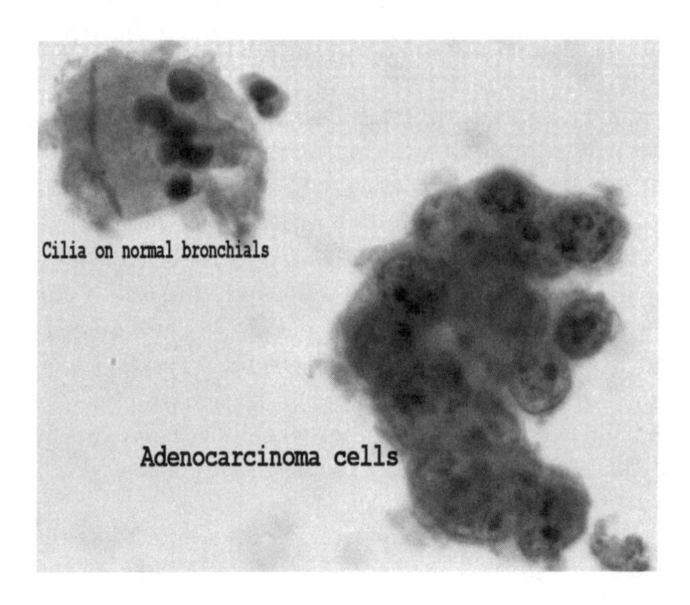

FIGURE 4.2 Fine needle aspirate of lung cancer. The normal epithelial cells are on the upper left. A clump of malignant cells, adenocarcinoma in this case, are on the lower right.

Notice that the malignant cells are crowded and lie on top of one another. The nuclei are large. Every malignant nucleus is much larger than any of the nuclei in the benign clump of epithelial cells. The chromatin, seen as blue specks in a nucleus, is dark, and shows no regular pattern of distribution from malignant cell to malignant cell.

4.4 Nuclear Size and Volume

The atypical nucleus is always larger than the nucleus of a normal, nondividing cell. The normal nucleus of a nondividing cell is about the same size as a mature red blood cell (about 7 to 9 microns in diameter). The nuclei of malignant cells can range from slightly larger than normal to over 50 microns in diameter. The difference in size between normal and malignant nuclei is one of the most dependable criteria for **malignancy**.

In addition to their increased nuclear size, cancer cells have greater nuclear size variation. A cancer may have one cell with a slightly enlarged nucleus adjacent to another cancer cell with a nucleus twice the size. A nonneoplastic population is composed of cells with very little variation in nuclear size among members of the population. Careful examination of a group of cells from a cancer almost always reveals many variations in nuclear size.

4.5 Nuclear Contour

Cancer nuclei are not as smooth or as round as nuclei of normal cells. They tend to have a more complex contour with jagged bumps and pits. The irregular shape of one cancer cell tends to be different from the irregular shapes of adjacent cancer cells.

4.6 Nuclear Texture

Nuclear texture relates to the distribution of chromatin. **Chromatin** is the material that stains blue with the standard **hematoxylin and eosin** stain used to colorize tissues mounted on glass slides. Chromatin is the mixture of DNA, RNA, and attached proteins that reside within the nucleus. Blue hematoxylin is preferentially bound to nuclear chromatin. Pink eosin is also bound to chromatin but to a lesser extent. In a normal cell, chromatin is evenly distributed, and the nucleus has a near-uniform distribution of chromatin. In a malignant cell, the distribution of chromatin varies greatly from place to place inside the nucleus, with large dense clumps of chromatin admixed with smaller clumps, separated by light areas where there seems to be no chromatin staining (parachromatin clearing). The distribution (texture) of chromatin in cancer cells is different from nucleus to nucleus.

4.7 Nucleolar Variability In Size, Shape, and Number

Each chromosome in a normal nucleus contains a nucleolar organizing region that is associated with a **nucleolus**, a circumscribed location within the nucleus in which RNA is synthesized and constructed into functional subunits. Although every nucleus has at least one nucleolus, the normal, quiescent cell contains between zero and six observable nucleoli, and these are typically small, round, discrete structures that stain pink (eosinophilic) with hematoxylin and eosin stain. Malignant cells tend to have more visible nucleoli than normal cells, and their nucleoli tend to vary greatly from cell to cell in size, shape, and number.

4.8 Cancerous Atypia and Reactive Atypia

Atypia is simply a change from the normal morphology of a cell. Whenever cells are exposed to a toxic agent, the cell responds with many morphologic changes that alter the appearance of the nucleus and that somewhat resemble the appearance of cancer cells. Cells with **reactive atypia** either die or recover without acquiring a genetic change that confers the **malignant phenotype** to daughter cells. The job of distinguishing reactive atypical cells from cancer cells often falls to the cytology laboratory.

The field of cytology deals with the study of cells, the basic living units of all organisms. Medical cytologists are highly trained microscopists who screen the specimens received in cytology departments. **Cytopathologists** are medical pathologists who have received additional, specialized training in cytology. In this book, medical cytologists and cytopathologists are referred to under the generic term *cytologist.* Cytologists can diagnose many pathologic processes, including cancer, by examining individual cells (rather than entire tissues).

Cytology specimens are different from tissue specimens. Tissue sections are thin slices of tissue that preserve normal tissue architecture and relationships among all the different cell types in the tissue. Cytology specimens consist of loose clumps or single cells that were scraped from a tissue surface (as in Papanicolaou smears of cervix) or from an internal tissue (as in a fine-needle aspiration), or were suspended in a body fluid (such as sputum or urine). About 50 million human cytology specimens are collected each year in U.S. cytology laboratories. The field of medical cytology is built on the premise that you can distinguish one pathologic process from another just by careful observation of individual cells.

4.9 Dysplastic Cells and How They Differ from Cancer Cells

The term **dysplasia** (Greek: dys = wrong, plasso = to mold, to shape) has several different medical meanings. Pediatricians use the term dysplasia to indicate a dysmorphic developmental disorder (i.e., a disorder in which the shape or size of the skeleton or some other observable body part is abnormal). This use of dysplasia is not related in any way to the neoplastic process. Dysplasia is also used to describe a growth insufficiency in newborns (bronchopulmonary dysplasia) and a type of growth defect in bone (fibrous dysplasia of bone). These uses of the word dysplasia are likewise unrelated to cancer.

Workers in the cancer field use dysplasia to describe a type of nuclear atypia that is passed from a cell to its progeny. Pathologists try to reserve the term dysplasia to describe early nuclear changes that characterize precancers (lesions preceding the development of invasive neoplasms). These early changes in precancer cells cannot always be distinguished from changes seen in cancerous cells, and, thus, both precancers and cancers are technically dysplastic.

Nuclear atypia present in precancer cells and cancer cells is the morphologic expression of accumulated genetic and epigenetic changes. Epigenetic changes occur in chromosomes and include structural changes, nuclear protein changes, and modifications of DNA bonded to carcinogens that do not involve alterations in the genetic code.

Most precancers can be "graded" by pathologists based primarily on the degree of nuclear atypia in the precancer as well as the presence of several other architectural features, such as the volume of normal cells replaced by precancer cells and the presence of abnormally shaped glands. The need to **grade** precancers is an indication that precancers progress over time to become more and more like the cancers into which they eventually develop (when **invasion** begins). The gradual increase of genetic abnormalities that occurs in precancers, as they advance in grade over time, has been well documented for the common precancers of humans (58–62).

4.10 Pathology of Common Precancers in Humans

Precancers are generally small lesions that are often composed of dysplastic cells that cannot invade or metastasize. There are exceptions (discussed in Chapter 5). In this chapter, we describe the general pathologic features of precancers and demonstrate the histopathologic features of the most common precancers: actinic keratosis, Barrett esophagus, cervical intraepithelial neoplasia, prostatic intraepithelial neoplasia, and colon adenoma.

4.11 Actinic Keratosis, the Typical Precancer

The typical precancer is a focal expansion of **atypical** cells growing in the preexisting scaffold of tissue from which it arises. As an example of a typical precancer, let us examine the **actinic keratosis** of skin.

Actinic keratoses are the most common precancers of humans. They appear almost exclusively on sun-exposed areas of the skin. After a certain age, all fair-skinned people who spend a great deal of time outdoors and who live in a sunny climate, develop multiple actinic keratoses, particularly on the face, neck, and arms. They usually occur as small patches of roughened (keratotic) skin or as reddened areas. These lesions are so common that that few of them are biopsied. It would be prohibitively expensive and of little practical benefit. Patients presenting to a dermatologist are spot-treated with liquid nitrogen wherever a suspicious lesion is seen or felt.

Actinic keratoses form in the **epidermis**, the outer surface of normal skin. The epidermis is produced by a dividing population of basal epidermal cells, located on the lower border of the epidermis. A basal epidermal cell divides to produce one post-**mitotic** cell (a fully differentiated epithelial, or squamous, cell that cannot divide), plus another basal cell, capable of repeated division. The nondividing squamous cells move upward through the epidermis, pushed by the successive nondividing cell produced by the next division of the underlying basal cell. With each division, the squamous cell rises higher and higher through layers of the epidermis. These nondividing cells become thinner and thinner as they ascend from the bottom layer to the top layer (Figure 4.3). Their nuclei become smaller and smaller, and their cytoplasm becomes highly keratinized (loaded with **keratin**, the same protein present in hair and fingernails). Eventually, the nondividing epidermal cell becomes an anucleate squame, filled with keratin, forming a thin cell resembling a pavement stone. After the epidermis has achieved a certain thickness, the topmost layer of keratinized cells sloughs into the environment. House-dust, the tiny particles that dance in sunbeams, is primarily sloughed squames.

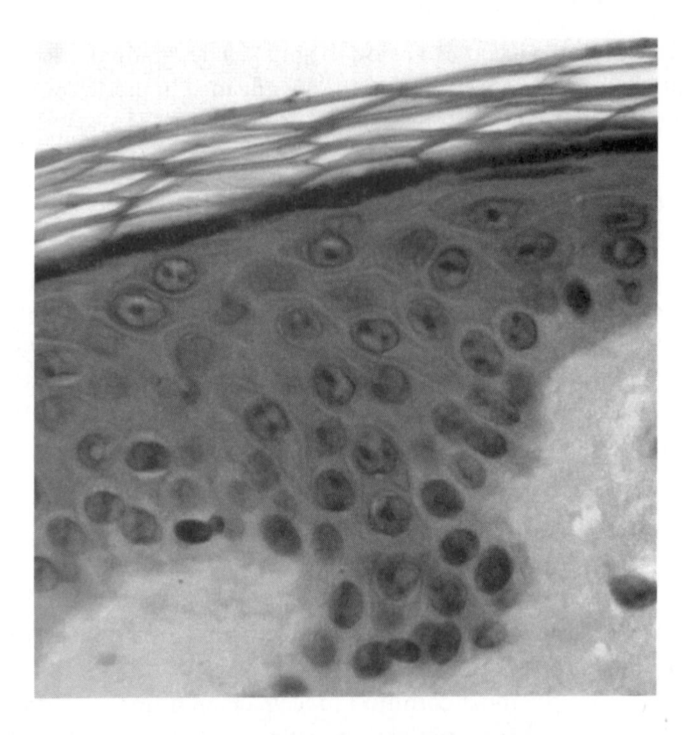

FIGURE 4.3 Every cell in the epidermis is generated from a dividing population confined to the lowest level. After division, cells rise through the epidermis, pushed upward by succeeding generations. Cells flatten (become squamous) as they move upward and eventually lose their nucleus. The topmost levels consist of flat, polygonal, keratin-filled squames.

Neoplasia is a disease of dividing cells, and the changes of precancerous dysplasia occur in the basal epidermal cells. The earliest morphologic changes of actinic keratosis are atypia of the basally located cells. Because cells of the upper levels of the epidermis are the nondividing progeny of basal epidermal cells, they retain atypia that is seen most convincingly in the lowest (basal) layer (Figure 4.4).

Over time, precancer can progress to fill the entire epidermis with highly atypical cells. In advanced precancer, the zone of proliferating cells is not confined to the lower levels. Mitosis is the stage of **cell division** in which complements of chromosomes separate and migrate into two daughter cells. In precancers with severe dysplasia, highly atypical cells and mitotic cells are present in every layer (Figure 4.5). When the entire epidermis is replaced by atypical cells, the lesion is called a **carcinoma** *in situ* (Latin: in place). This term signifies that the precancer has progressed to a state where it has all the morphologic properties of a cancer with the exception of invasion. All epithelial precancers develop in the same manner, regardless of cause or tissue of origin (List 4.11.1). As far as we can tell, all the steps in precancer development are capable of being reversed or stabilized. In other words, the **progression** of precancers to cancer is not inevitable.

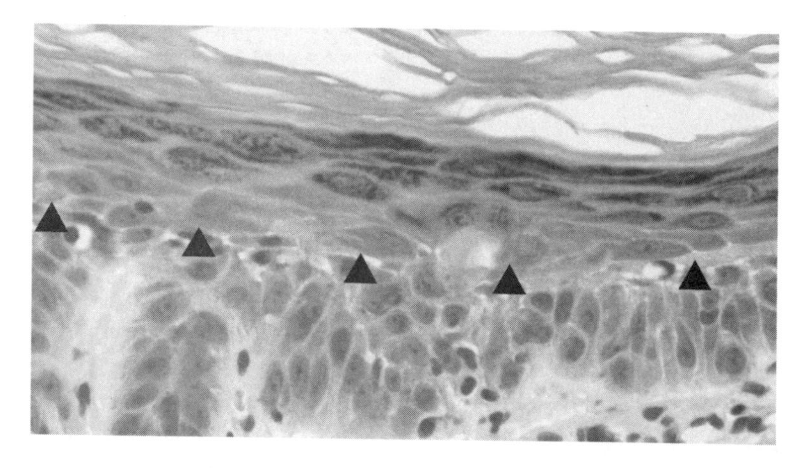

FIGURE 4.4 Actinic keratosis. There is a demarcation (line of triangles) between atypical cells in the basal layers of the epidermis and normal-appearing cells of the upper layers. The atypical cells have enlarged, irregular nuclei.

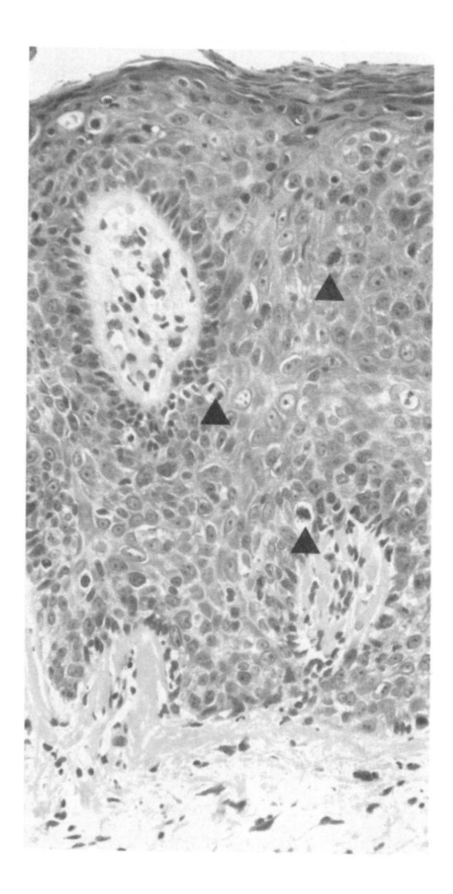

FIGURE 4.5 Squamous cell carcinoma *in situ* (Bowen disease of skin). Triangles point to cells in mitosis. These mitotic cells are not confined to the basal layer of the epidermis.

List 4.11.1 Morphologic steps in precancer development.

1. Focus of dysplastic cells appears in the layer of epithelium (squamous or glandular) reserved for dividing cells.
2. Nuclear atypia present in dividing cells is retained in the dividing progeny of dysplastic cells and, to a varying extent, in the nondividing progeny of dysplastic cells. In the earliest stages of dysplasia, this results in a mucosa with a zone of atypia at the basal level and a zone of somewhat less atypia in the layers of nondividing progeny cells.
3. Over time, the degree of atypia in the basal cells becomes more extreme. Likewise, the degree of atypia in the nondividing progeny of the dysplastic cells also worsens. The zone of dividing cells expands, and dividing cells are present in areas other than the normal, anatomically confined, dividing zone of the epithelium. In the case of a squamous epithelium, this means that dysplastic cells are present in layers of the epithelium above the basal layer. In the case of glandular mucosa composed of crypts or pits (such as the stomach and intestines), this means that the zone of cell division has expanded from the normal dividing zone in the bottom of the crypt to otherwise nondividing zones at higher levels in the crypt. In the case of glands that form acini, this means that the dividing, dysplastic cells have spread through the acinar connections, into adjacent acini.
4. Eventually the zone of dividing dysplastic cells expands to include the entire epithelium. At this point, the precancer is termed a "carcinoma *in situ*," indicating that a focus of epithelium (e.g., squamous lining, focus of crypts, portion of a duct, a collection of adjacent ducts, or patch of glands) is completely populated by dysplastic cells that have replaced the normal cells. In carcinoma *in situ*, there is no invasion into adjacent tissues. The dysplasia is still in its proper place (i.e., "*in situ*").

4.12 Barrett Esophagus

The esophagus is a tube through which food passes from the mouth to the stomach. Once in the stomach, food is bathed in acid produced by the stomach. The stomach has a special lining, capable of producing stomach acid yet resisting the toxic effects of the acid it produces. Consequently, the stomach is a complex organ lined by several types of cuboidal, glandular cells in pits (Figure 4.6).

Under normal conditions, there is a **sphincter** at the lower end of the esophagus, composed of circumferential muscle inside the esophageal wall. The purpose of the esophageal sphincter is to stop stomach contents from flowing backward into the esophagus. If the lower esophageal sphincter becomes incompetent, then acid flows into the esophagus and injures the squamous **mucosa**. This is called **gastroesophageal reflux disease (GERD)**. Many causes can lead to esophageal sphincter incompetence, but the most common cause is normal aging. GERD is a common problem and accounts for most esophageal disease in Americans. The esophagus is lined by squamous cells, the same epithelium present in the oral cavity (Figure 4.7).

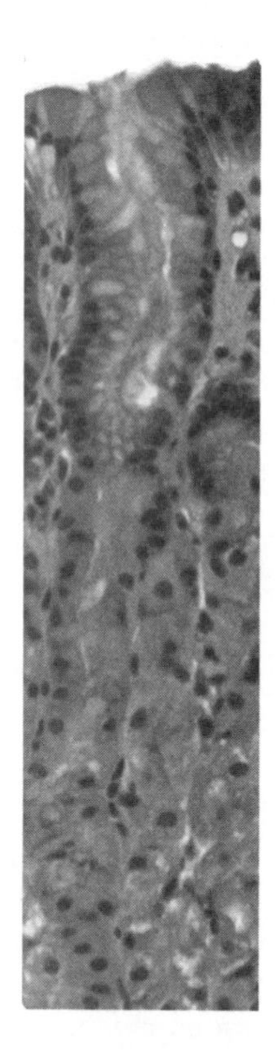

FIGURE 4.6 Normal body of stomach. The stomach is lined by gastric pits. The upper levels of the pit (isthmus) are populated by a single cell type, the gastric enterocyte, which forms a layer of cells, each cell containing a large drop of mucus (clear zone). The lower levels of the pit (neck and base) have several different specialized cell types, that produce hydrochloric acid (parietal cells), digestive enzymes (chief or **zymogen** cells), and intestinal hormones (**enteric endocrine** cells).

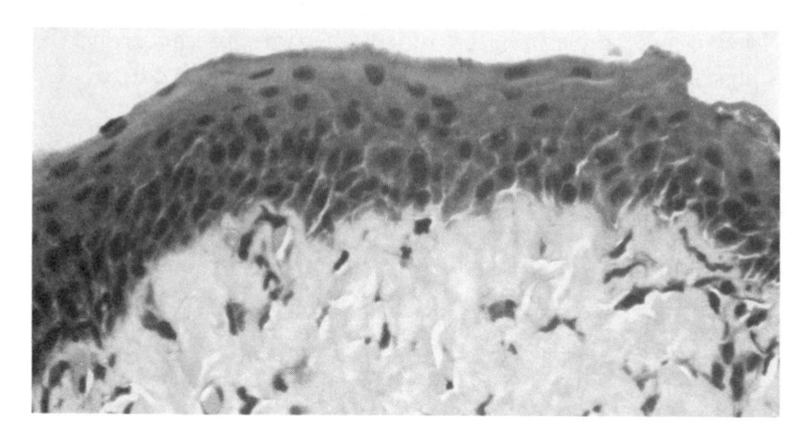

FIGURE 4.7 Normal esophagus, lined by squamous mucosa (top), with underlying submucosa (bottom).

When the esophagus is chronically injured by GERD, the squamous lining may be focally replaced by glandular lining (Figure 4.8). This phenomenon is called **intestinal metaplasia** or Barrett esophagus. **Metaplasia** is a process whereby a normal tissue is

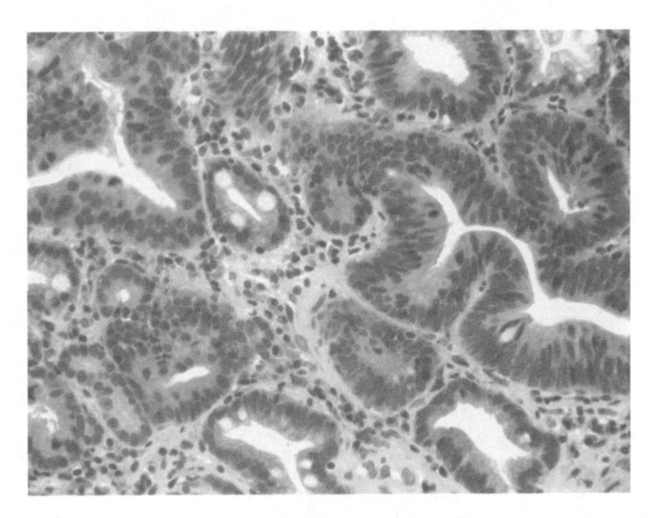

FIGURE 4.8 Barrett esophagus. Intestinal metaplasia without dysplasia. The squamous lining is focally replaced by glands. The glands are populated by enterocytes that have the same morphologic features as intestinal enterocytes.

replaced by a tissue with properties of an identifiable tissue from some other anatomic site. The theory is that as GERD worsens, the body replaces the normal esophageal squamous epithelium with the **epithelium** of another tissue possibly better suited to withstand a hydrochloric acid bath.

If GERD worsens, then over time the physiologic defenses of the esophageal mucosa (squamous regeneration and intestinal metaplasia) may fail to compensate for the damage. The patient may experience pain (heartburn). The esophagus may ulcerate and bleed. In rare cases, the esophageal mucosa becomes precancerous. Occasionally, dysplasia is present within a focus of Barrett esophagus (Figure 4.9).

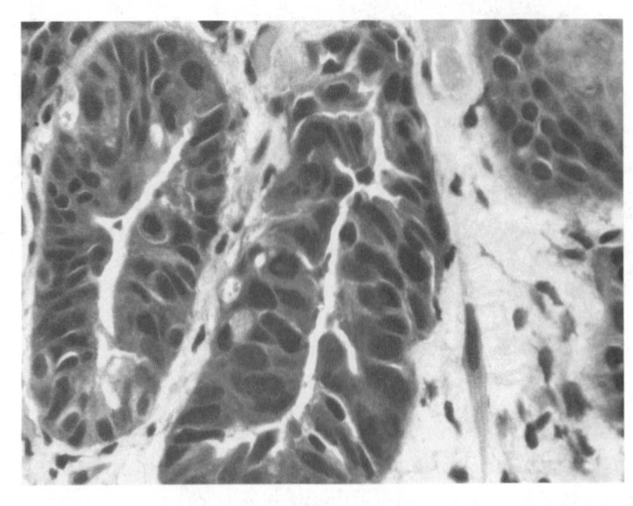

FIGURE 4.9 Barrett esophagus with high-grade dysplasia. The cell nuclei are large, crowded, dark, and atypical.

Barrett esophagus with dysplasia may eventually progress to esophageal adenocarcinoma. Similarly, squamous mucosa of the esophagus may develop dysplasia, which can progress to esophageal squamous cell carcinoma. Esophageal carcinoma (squamous or glandular) has a very poor prognosis.

4.13 Cervical Intraepithelial Neoplasia

The uterine cervix (Latin: *uterus* = womb, *cervix* = neck) is lined by squamous epithelium (the **ectocervix**) and glandular mucosa (the **endocervix**). The endocervix is located in the central part of the cervix, at the os (Latin: *os* = mouth) that opens into the uterine body. The margin, or "transformation zone," between the endocervix and ectocervix is the site where cervical carcinoma and its precancerous lesions are most likely to occur.

Abnormalities in cervical cells are detected on Pap smears. Pap smears are obtained during routine gynecologic examination. The gynecologist inserts a brush, swab, or wooden spatula through the vagina to the endocervical os and gently scrapes off some of the cells. The cells are either smeared directly onto a glass slide and immediately fixed in alcohol or dispersed in a specimen jar containing alcohol and later spread onto a glass slide. The cells are stained and examined by a cytologist. An adequate Pap smear contains endocervical (glandular) and ectocervical (squamous) cells (Figure 4.10).

Cervical intraepithelial neoplasia is atypical cervical squamous epithelium. As in all epithelial precancers, atypia begins in the basal layer, the layer of dividing cells located at the bottom of the epithelial lining. Atypical cells rise through the cervical epithelium until they reach the top-most level. Scraping the cervical epithelium removes cells from the upper layers. The **nuclear atypia** persists in cells present on the Pap smear (Figure 4.11).

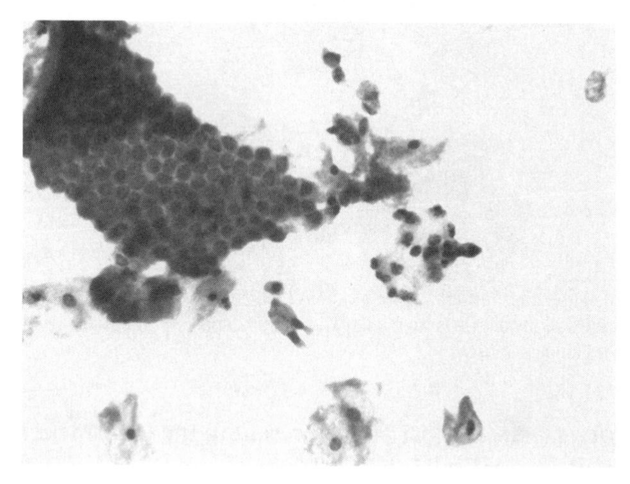

FIGURE 4.10 Normal gynecologic (Pap) smear. A large clump of endocervical cells is shown in the upper left corner. A few scattered, flat squamous cells are also seen on the right and on the bottom.

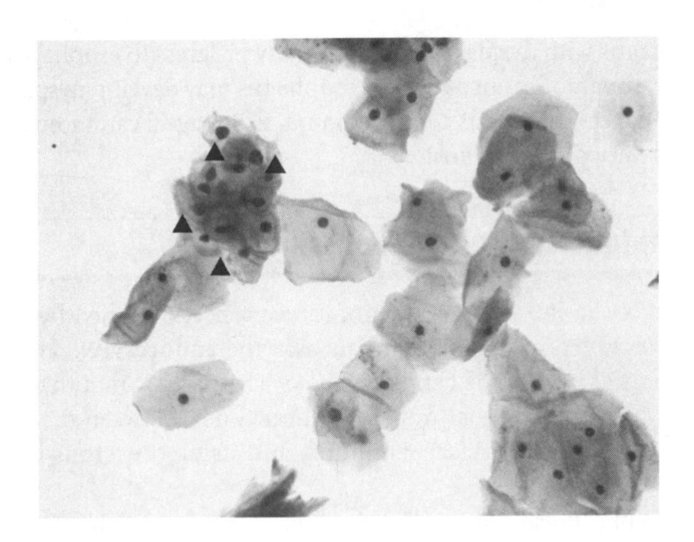

FIGURE 4.11 Low-grade squamous intraepithelial lesion. Most cells are normal squamous cells. A small clump of cells (enclosed by triangles) shows mild nuclear atypia. The nuclei are somewhat enlarged, dark, and angular, with perinuclear clearing of the surrounding cytoplasm. These features of low-grade squamous intraepithelial lesion correspond to mild cervical dysplasia on tissue biopsy.

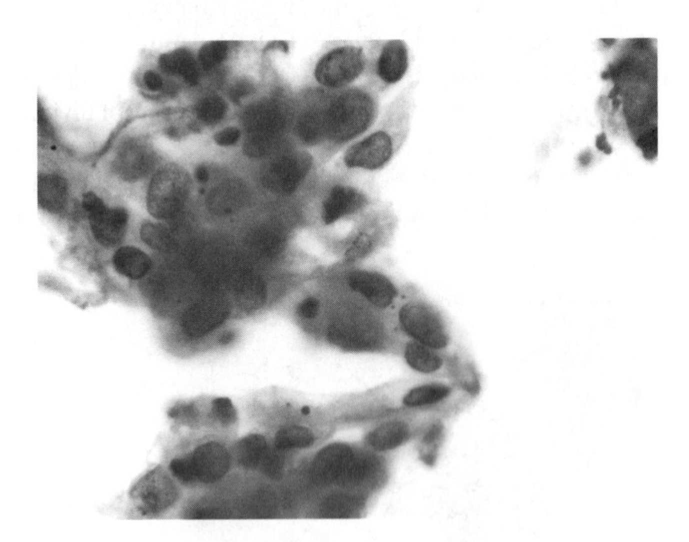

FIGURE 4.12 High-grade squamous intraepithelial lesion. All the cells are abnormal. The nuclei are large, dark, and vary in size and shape from cell to cell. Cells from a high grade squamous intraepithelial lesion correspond to moderate to severe dysplasia on cervical biopsy.

As dysplasia worsens, the nuclear atypia present in the Pap smear worsens. In many cases, the number of atypical cells also increases (Figure 4.12). A tissue biopsy corresponding to a high-grade intraepithelial lesion is shown in Figure 4.13.

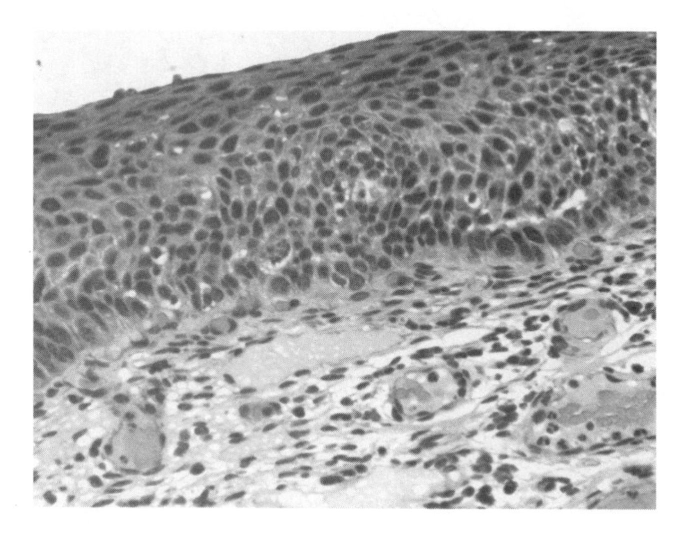

FIGURE 4.13 Tissue biopsy of cervix, corresponding to high-grade squamous intraepithelial lesion on Pap smear.

As cervical precancer worsens, the full thickness of epithelium is replaced by highly atypical cells. Cell division is no longer restricted to the basal layer and is present throughout the epithelium. At this point, the lesion is a squamous cell carcinoma *in situ*. Many of these lesions progress to invasive cervical carcinoma.

The purpose of Pap smear screening is to detect cervical neoplasia while it is still in its precancerous stage. Cervical precancer can be removed by a relatively simple gynecologic procedure. Once the precancer is removed entirely, it cannot progress to cancer.

4.14 Prostatic Intraepithelial Neoplasia

One of the most common cancers occurring in men is prostate cancer. The risk of developing prostate cancer increases with age. By the age 80 to 90 years, 70% to 90% of men have prostate cancer confirmed at autopsy (21, 22). Fortunately, most prostate cancer is indolent and does not cause significant **morbidity** or mortality for the majority of the men who have it. Nonetheless, about 30,000 American men die each year from prostate cancer. The prostate gland is composed of glands within muscle and connective tissue (Figure 4.14).

The precancer for prostate cancer is **prostatic intraepithelial neoplasia (PIN)** (21, 63). As one would expect, PIN occurs in an earlier age group than prostate carcinoma. It is not unusual to find PIN in men in their 20s and 30s (64). Prostatic carcinoma occurs at a higher rate in African-American men than in men of other ethnicities. Not surprisingly, African-American men have high rates of PIN. When PIN is present in African-American men, it is likely to show more extensive spread in the prostate than PIN present in men of other ethnicities (64).

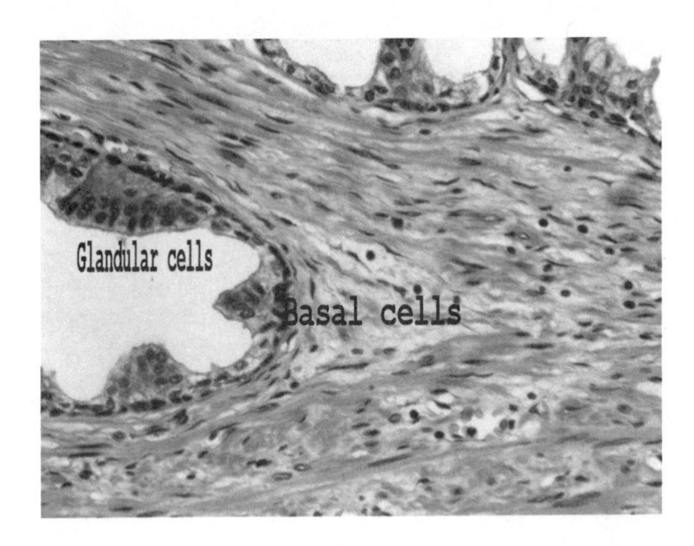

FIGURE 4.14 Normal prostate tissue. There are several glands, each containing a lumen (empty space) lined by epithelial cells. Epithelial cells occupy the glandular and basal layers. The glandular cells are round with relatively clear cytoplasm, and extend into the lumen. The basal cells are small, dark, and form a single-cell layer between the glandular cells and the surrounding fibromuscular tissue.

PIN occurs in the glands of the prostate (Figure 4.15). Nuclei of PIN cells are larger, darker, more irregular, and more crowded than nuclei of normal prostate glandular cells. The normal-appearing basal cell layer is preserved

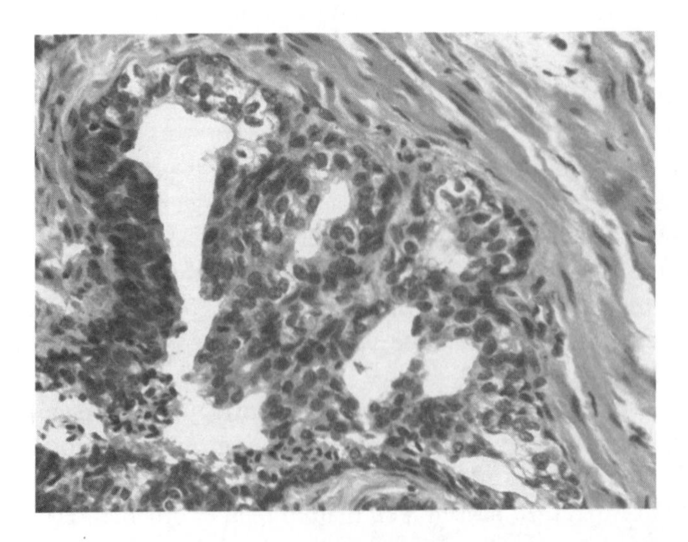

FIGURE 4.15 High-grade prostatic intraepithelial neoplasia (PIN). There are large, hyperchromatic nuclei with prominent nucleoli. The nuclear have lost their basal orientation and have become crowded with overlapping.

Viewed at higher magnification, the cells of high-grade PIN fill most of the glandular lumen, with multilayered, crowded cells, with some atypia in most cells, primarily characterized by a large, dark nucleus (Figure 4.16). Most nuclei have a discernible, somewhat enlarged nucleolus. The gland lumen is traversed in places by cells with the appearance of glands within the gland. No atypical cells are present in the fibromuscular tissue.

Prostatic adenocarcinoma has glands apparently streaming through the fibromuscular tissue (Figure 4.17). This is invasion. Notice that the basal cell layer that surrounds normal glands and PIN glands is absent.

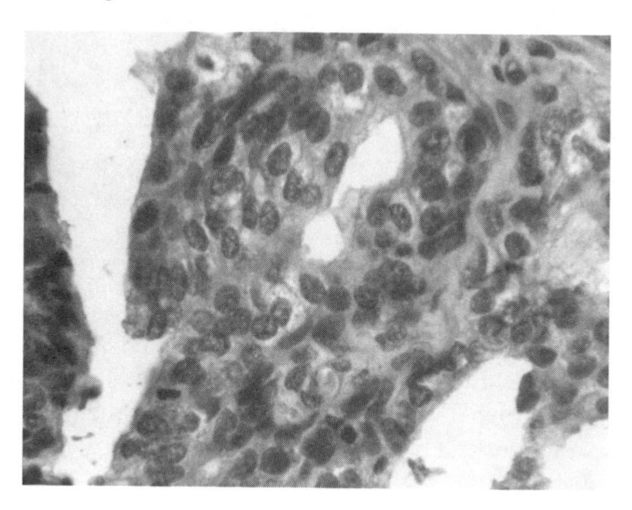

FIGURE 4.16 High-grade PIN, higher power.

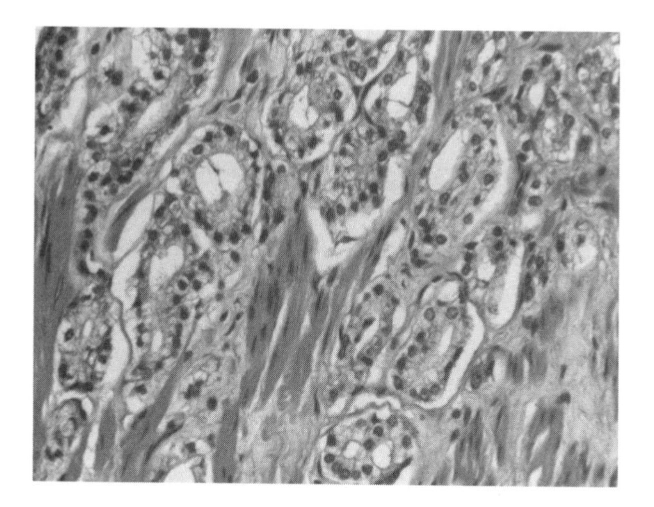

FIGURE 4.17 Prostatic adenocarcinoma. There are small, infiltrating, back-to-back glands with enlarged, hyperchromatic nuclei showing prominent nucleoli.

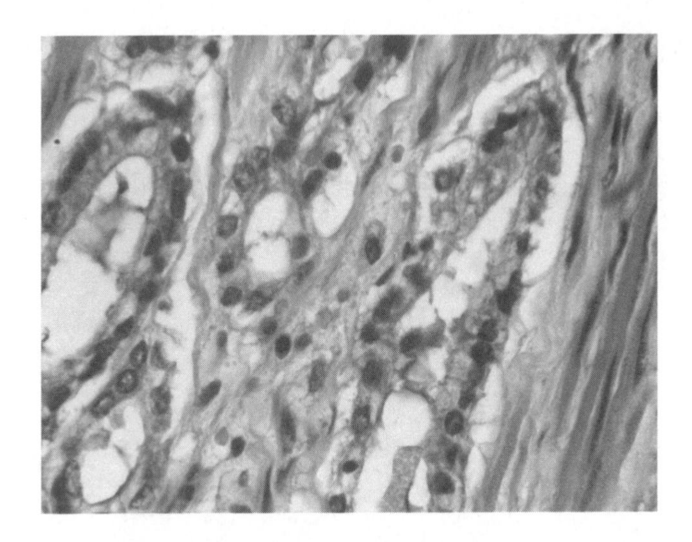

FIGURE 4.18 Prostatic adenocarcinoma, higher power.

Viewed at higher magnification, we see that the malignant glands have different shapes and sizes (Figure 4.18). Some glands abut and crowd other glands. The nuclei of the malignant cells have highly variable sizes and shapes. Some nuclei are dark. Other nuclei are somewhat lighter and have irregular patches of lightness and darkness scattered throughout the nucleus.

4.15 Colon Adenoma

The **colon** is the last portion of the intestine. The colon has a uniform histologic appearance, consisting of long, straight glands (crypts), often said to resemble test tubes. Each colon crypt is lined by a homogeneous population of cells dominated by mucus-filled intestinal enterocytes, or goblet cells (Figure 4.19).

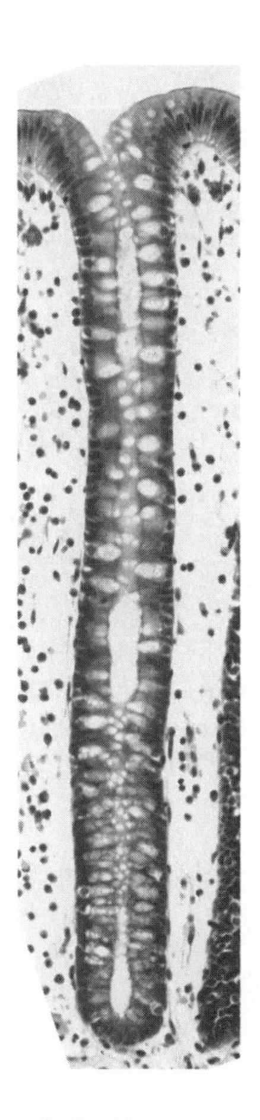

FIGURE 4.19 Normal colonic mucosa. The crypt is lined by clear, mucus-producing (goblet) cells. Outside the crypt is the lamina propria, composed of loose connective tissue and scattered hematolymphoid cells. There are prominent lymphocytes in the lamina propria.

Adenocarcinoma of the colon is the second most common cause of cancer deaths in the United States. The common precursor for adenocarcinoma of the colon is the colon adenoma. The adenoma is populated by infolded glands (Figure 4.20). Adenoma cells have less mucus than the normal colon enterocytes. Their nuclei are somewhat larger than normal nuclei (Figure 4.21).

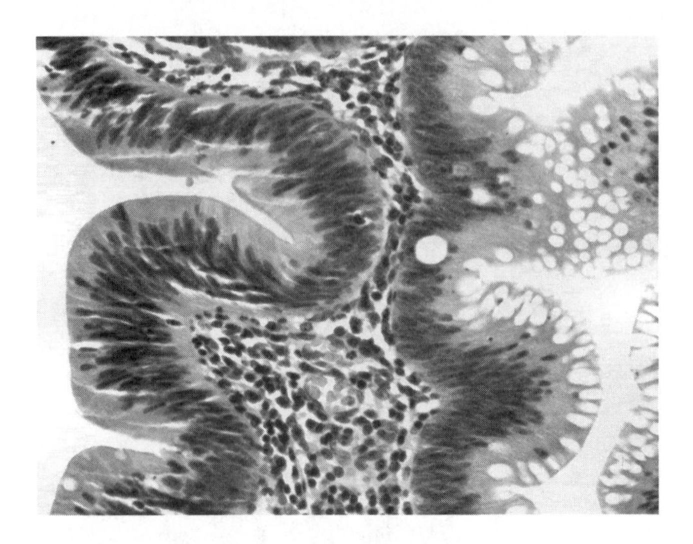

FIGURE 4.20 Histology of colon adenoma. The basal cell nuclei of the colonic epithelium are crowded and hyperchromatic, with a heaped-up appearance. Mucin production is variable, ranging between slightly decreased (right) and markedly decreased (left). The basement membrane is intact.

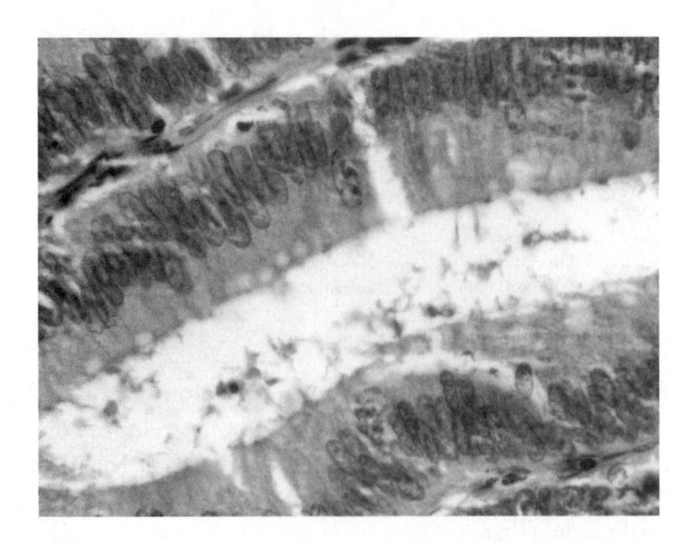

FIGURE 4.21 Colon adenoma, higher power. Neoplastic glands are lined by a single row of cells, but the cells are highly crowded and nuclei are distorted by their neighbors. The size of each nucleus in the adenoma is homogeneous, but comparison with nuclei of normal cells in the lamina propria (connective tissue between the glands) shows that the neoplastic cells are greatly enlarged. Mucus in neoplastic cells is reduced compared to normal and appears as small droplets near the apex of cells.

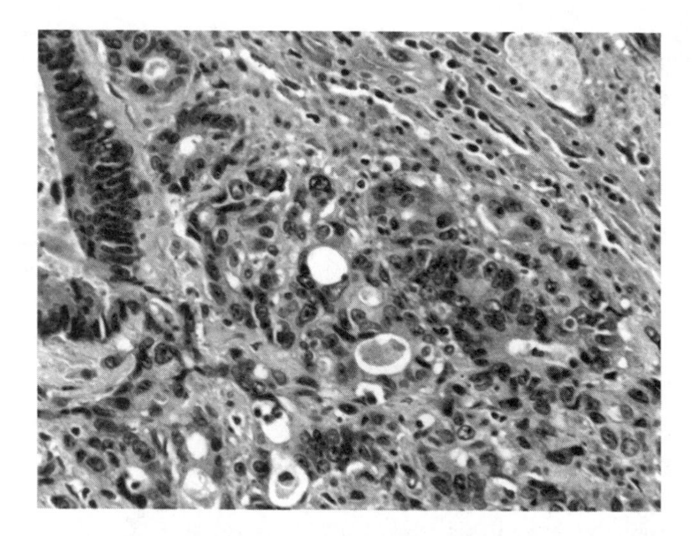

FIGURE 4.22 Colon adenocarcinoma, invading into the muscularis propria (deep muscle layer of the colon). There are small glands, with variability in gland size and shape, and complex folding of the epithelium. Nuclei are enlarged and hyperchromatic. There is a fibrous reaction (desmoplasia) around the invading, cancerous glands.

The defining property that distinguishes a **colon adenoma** from a colonic adenocarcinoma is invasion (Figure 4.22). Aside from invasion, many morphologic features are more commonly present in cancer than in adenomatous glands. Nuclear atypia is usually more severe in cancer. Tubules of the adenoma are often replaced by small glands. Invading clusters of glands, or single glands, or even single malignant cells of the carcinoma, often evoke a reactive growth of connective tissue and inflammatory cells. Reactive growth of nonneoplastic connective tissue around invading tumor cells is called **desmoplasia**, a feature that is consistently absent from colon adenomas. As colon adenomas grow larger, the risk of cancer increases. Finding a focus of invasive carcinoma within a large colon adenoma is not unusual. This suggests that a subclone of the adenoma has transformed into cancer.

4.16 Common Precancers

There are hundreds of different precancers of humans (65). List 4.16.1 shows some of the most important, well-studied precancers. A more extensive listing of precancers is included in the Appendix, and descriptions of many of the best understood precancers are included in the Glossary.

List 4.16.1 Examples of common precancers and cancers into which they develop (66).

Squamous dysplasias
- Actinic keratosis → squamous cell carcinoma of skin
- Squamous dysplasia of esophagus → squamous cell carcinoma of esophagus
- Squamous dysplasia of bronchus → bronchogenic carcinoma
- Squamous intraepithelial lesion of cervix → squamous cell carcinoma of cervix

Glandular dysplasias
- Prostatic intraepithelial neoplasia → prostatic adenocarcinoma
- Barrett esophagus with dysplasia → adenocarcinoma of esophagus
- **Gastric dysplasia** → adenocarcinoma of stomach
- Pancreatic intraepithelial neoplasia → ductal carcinoma of pancreas (67)

Other dysplasias
- **Ovarian intraepithelial neoplasia** → ovarian carcinoma (68)
- Myelodysplasia → acute myelogenous leukemia (69)
- **Monoclonal gammopathy of undetermined significance** → **multiple myeloma** (70)
- Dysplastic nevus → melanoma (71)

Chapter 4 Summary

Precancers are the earliest morphologically discernible lesions that precede the development of invasive cancer. The histologic hallmark of precancers is **cytologic** dysplasia, the morphologic manifestations of genetic and epigenetic alterations in the nucleus. The features of dysplasia include irregularities of nuclear size (enlargement) and shape, distribution of chromatin, and irregularities of shape, size, and number of nucleoli. Over time, these changes increase within individual cells and in the number of involved cells. In epithelial tissues delimited by **basement membrane**, the typical progress of precancer development involves expansion of the dysplastic cell population, beginning with the most basal (lowest) epithelial cells and expanding into the full thickness of the epithelium. The expansion of dysplastic cells confined by a basement membrane is called intraepithelial neoplasia. After normal epithelium is replaced by dysplastic cells, invasion may occur. Invasion marks the end of the precancer stage of neoplastic development and the beginning of the cancer stage.

CHAPTER

Biology of Precancers

5

The beginning is the most important part of the work.

—Plato

5.1 Background

One of the problems with studying precancers is that the term *precancer* is often confused with many unrelated terms, particularly ***early cancer*** or *small cancer.* In addition, there are many near-synonyms for precancer including premalignancies, premalignant lesions, preneoplasias, preneoplastic lesions, incipient cancers, incipient neoplasia, incipient neoplasms, intraepithelial neoplasias (IEN), cancer precursors, pretumors, precancerous lesions, precancerous states, and preinvasive cancers. The plethora of terms reflects the difficulty of choosing a "best" term for the precancerous lesions. Currently, the term ***intraepithelial neoplasia (IEN)*** enjoys wide usage in the community of pathologists.

Most cancers arise from epithelial surfaces. An epithelial surface is a flat, tubular, or round surface lined by polygonal cells (epithelial cells), and demarcated from the subjacent **lamina propria** (loose fibrous tissue) by a thin, underlying (basement) membrane composed of complex proteins and carbohydrates (Figure 5.1).

Examples of epithelial cancers are squamous cell carcinoma, bronchogenic carcinoma, prostatic carcinoma, ductal carcinoma of breast, and serous carcinoma of the ovary. Precancers for common epithelial cancers grow as a noninvasive clonal collection of atypical cells localized to the epithelial surface. These precancers are referred to as intraepithelial neoplasias or intraepithelial neoplasms.

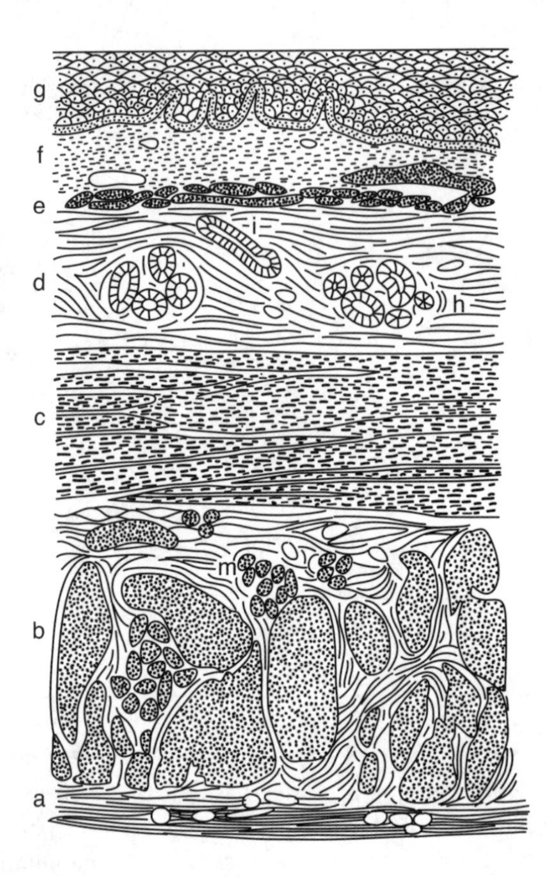

FIGURE 5.1 Cross-section of a tubular or spherical internal organ (esophagus, stomach, bowel, gallbladder, urinary bladder). The top layer is the epithelium. Precancers that develop from the epithelium remain confined to the epithelial layer, as long as they are precancers. During this precancerous phase, the epithelial-derived precancers are aptly named intraepithelial neoplasias. When invasion takes place, tumor cells move into the underlying lamina propria or beyond. Once this occurs, the lesions are no longer precancers; they are invasive cancers.

Key to figure: g = epithelial layer, collection of epithelial cells delineated from the underlining lamina propria by a thin, basement membrane; f = lamina propria; e = (superficial) muscularis mucosae; d = submucosa; c,b = two layers of (deep) muscularis propria; the inner layer (c) is circular; the outer layer (b) is longitudinal; a = serosa. Adapted from Gray, Henry. *Anatomy of the Human Body*. Lea & Febiger Publishers, 1918.

The term *intraepithelial neoplasia* has limitations (List 5.1.1).

Likewise, the term *preinvasive cancer* raises an existential question. Use of the term *preinvasive cancer* implies that precancers have achieved the biological properties of a cancer. This assumption may not be true. Precancers may lack constitutive properties of cancer or may have certain attributes that are absent in cancers. At this point, there is insufficient knowledge to conclude that precancers are types of cancer. In this book, we use the term *precancer* because it includes all known lesions from which malignancies develop.

List 5.1.1 Limitations of the term *intraepithelial neoplasia.*

1. Not all epithelial precancers are intraepithelial (i.e., confined to an epithelium). Many epithelial dysplasias have a well-defined territory bounded by a thin membrane (the so-called basement membrane) inserted between the epithelium and the underlying stroma. But not all premalignant epithelial lesions can be identified by the presence of atypical cell populations delimited by a basement membrane. Dysplastic lesions of the liver, kidney, thyroid, and adrenal gland are not delimited by a basement membrane (72).
2. Not all precancers are epithelial (72). Intratubular germ cell neoplasms of testis, myelodysplasias, and nonautonomous lymphomas are examples of nonepithelial precancers.
3. Not all intraepithelial neoplasms are precancers. Neoplasms that are intraepithelial but not precancers include seborrheic keratoses, intraepidermal nevi, and common warts.

The names and the definitions we assign to objects forge our perceptions. If we think of stars as pinpoint holes in the fabric of space where heavenly light pours into the sky, then our concept of ourselves, our universe, and our purpose in the universe is fundamentally different from concepts held by those who think of stars as exploding balls of gas. This chapter is about the precise definition of precancer. We cannot make progress in the field until all participants share a common understanding of precancer biology. You can think of this chapter as an introduction to the remarkably fussy field of medical nomenclature.

5.2 Formal Definition of Precancer

An NCI workshop was convened on November 8–9, 2004, at George Washington University Medical Center. A definition of precancers was developed over two days of workshop discussions (73). Attendees included Stanley R. Hamilton, Jorge Albores-Saavedra, Ronald A. DeLellis, David L. Page, John N. Eble, Ralph H. Hruban, William D. Travis, David G. Bostwick, George L. Mutter, and Donald E. Henson. The committee concluded that five defining properties must all apply to putative precancers (List 5.2.1). This definition is a departure from prior definitions, which were based on histopathologic features (e.g., confinement of the lesion within a basement membrane and nuclear atypia).

List 5.2.1 The five defining properties of precancer.

1. Evidence must exist that the precancer is associated with an increased risk of cancer.
2. A precancer must have some chance of progressing to cancer, and the resulting cancer must arise from cells of the precancer.
3. A precancer must be different from the normal tissue from which it arises.
4. A precancer must be different from the cancer into which it develops. A precancer may have some, but not all, of the molecular and phenotypic properties that characterize the cancer.
5. There must be a method by which the precancer can be diagnosed.

5.3 Precancer Properties

Precancers have properties that are different from properties of cancer (List 5.3.1). Let us examine each property of the precancers.

1. **Precancers can be defined and studied as biological entities.**
   - Corollary: Properties of precancers are different from properties of normal cells and of cancer cells.

Although there is much that we need to learn about the precancers, we know enough to define the precancers and to distinguish them from cancers. Pathologists can review biopsies of precancers and distinguish them from normal tissues and from cancers by their histologic features. Precancers are among the most common specimens received in surgical pathology departments, and the most common precancer specimens are actinic keratoses, **dysplastic nevi**, prostatic intraepithelial neoplasias, cervical intraepithelial neoplasias, and colon adenomas. Hundreds of precancers have been well-described (66). If precancers can be described, then they can be diagnosed. If they can be diagnosed, then they can be studied by cytogeneticists, molecular biologists, and cell biologists.

2. **The incidence of precancers is higher than the incidence of cancers.**
   - Corollary: Many precancers regress or fail to progress.
   - Corollary: The incidence of low-grade precancers is higher than the incidence of high-grade precancers.

Because every cancer arises from a precancer, although not all precancers lead to cancers, the number of precancers must exceed the number of cancers. The earlier the precancer (i.e., the less advanced the precancer), the greater the likelihood of regression; therefore, the incidence of low-grade precancers is higher than the incidence of high-grade precancers.

List 5.3.1 Precancer properties.

1. Precancers can be defined and studied as biological entities.
   - Corollary: The properties of precancers are different from the properties of normal cells and of cancer cells.
2. The incidence of precancers is higher than the incidence of cancers.
   - Corollary: Many precancers regress or fail to progress.
   - Corollary: The incidence of low-grade precancers is higher than the incidence of high-grade precancers.
3. On average, precancers develop in younger individuals than do cancers.
4. Every cancer has a precancer.
   - Corollary: If all precancers were eradicated, there would be no cancers.
5. Treating a precancer is easier than treating a cancer.
6. A single precancer may develop into any one of several closely related cancers.
   - Corollary: There are more types of cancers than there are types of precancers.
7. Precancers contain a characteristic genetic abnormality that distinguish one class of cancer from another class of cancer.
   - Corollary: Agents that target an essential pathway in a particular cancer, also target the same pathway in the precancer for its corresponding cancer and for all closely related cancers.
   - Corollary: Precancers lack uncharacteristic genetic abnormalities that are accumulated by cancers during tumor progression.
8. Precancers can be distinguished from cancers into which they develop.
9. Precancers can be sensibly classified into classes of lesions with shared biological properties.
10. Precancers that progress do so into cancers and into no other type of lesion.
11. All agents that cause cancers also cause precancers.
12. Cancer that develops from a precancer must be the clonal descendant of a precancer cell.
    - Corollary: Cancers arise from precancers; often the precancer is seen in the same location where the cancer arises.
    - Corollary: Cancers that arise from precancers often have characterizing morphologic features present in the precancer (74).
    - Corollary: Cancers carry the genetic alterations of their precancer.

In a long-term follow-up study of 894 women with moderate dysplasia of the cervix, Nasiell and coworkers found that regression occurred in about half the cases (54%) (75). **Persistence** was present in 16% of lesions. Progression to more advanced dysplasia occurred in 30% of lesions. Scrutiny of dysplastic lesions may indicate which lesions are most likely to progress. Fu and coworkers found that dysplastic lesions that were **euploid** or polyploid (had a normal number of chromosomes or some integer multiple of the normal number of chromosomes), accounted for 85% of the regressing dysplasias. Nonregressing lesions were almost all aneuploid (i.e., had an abnormal number of chromosomes) (76).

In animal models, carcinogenic protocols tend to produce multiple precancers. Over time, some of these precancers develop into cancers (77–79). Multiple occurrence of precancers is also seen in humans. An individual who has hundreds of actinic keratoses is likely to have a small number (perhaps zero) of squamous cell carcinomas. An individual with hundreds of **nevi** will likely have a smaller number of atypical nevi and a very small number of malignant **melanomas** (80). Colon adenomas are often synchronous and multiple, but it is rare to find patients with multiple colon carcinomas. List 5.3.2 outlines why most precancers regress.

Regression is a phenomenon that is not seen in large epithelial cancers. After a precancer has progressed to a cancer (i.e., has become invasive), it will, in almost all cases, continue to grow, invade, and spread. Regression is a common phenomenon among lymphomas (89) and has been reported in as many of 30% of untreated follicular lymphomas (90). Morphologically distinctive precancers have not been described for lymphomas (i.e., there are no lesions recognized as **prelymphomas**). It may be that among lymphomas, there is a biologically distinct group of lesions that are prelymphomas.

List 5.3.2 Possible mechanisms of precancer regression.

1. The population of precancer cells may not have acquired all the biological properties necessary for sustained growth. These properties include **angiogenesis**, heightened **glycolysis** (to survive in a **hypoxic** microenvironment), invasiveness, etc.
2. The population of precancer cells may be unlucky. Monte Carlo simulations of tumor growth have shown that small populations often die out when the cell growth rate and the cell death rate of a neoplastic population are of similar magnitudes (81, 82).
3. The population of precancer cells is sufficiently **immunogenic** to stimulate a host response that eliminates the lesion (sometimes seen in early melanomas and nevi and in immunogenically stimulated urothelial cell tumors of bladder (83, 84).
4. The tumor differentiates into a nondividing population of cells (putative mechanism for regression in **keratoacanthoma**) or into a mature population of benign cells (85, 86). Alternatively, the tumor cells may fail to immortalize (i.e., cannot achieve unlimited cell replication) (87).
5. The growth advantage of the precancer, compared to normal cells, is lost when a growth stimulus is removed from the host environment (e.g., hormone-responsive tumors, and some cases of ***Helicobacter***-induced MALTomas [mucosa associated lymphoid tissue lymphomas]).
6. Virally induced precancers for which viral replication ceases (e.g., tumors occurring and regressing in transiently immunosuppressed patients) (88).

Obviously, if we could induce tumors to follow a naturally occurring regression pathway, we could eliminate tumors. The mechanisms of tumor regression are a vastly underexamined area of cancer research.

3. **On average, precancers develop in younger individuals than do cancers.**

Precancers progress over time into cancers. This means that for any given precancer, the average age of individuals in whom the precancer occurs must be younger than the average age of individuals in which the developed cancer occurs.

Ductal breast cancer is thought to develop through a series of morphologically distinct precursors: **Intraductal hyperplasia (IDH) → atypical intraductal hyperplasia (AIDH)** → intraductal carcinoma, also known as ductal carcinoma *in situ* (DCIS) → invasive breast carcinoma. The published ages for these different lesions seem to bear out a chronological precedence for precancers (List 5.3.3). Similar observations have been noted for the age of occurrence of cervical intraepithelial neoplasia and cervical cancer. The NCI's Surveillance, Epidemiology and End Results (SEER) program provides deidentified data on most malignant neoplams occurring in U.S. citizens. The SEER public data files covering years 1973 to 2005 also includes data on cervical precancers (10). Analysis of the SEER data yields the following summary data for the average age of occurrence of cervical *in situ* carcinomas, microinvasive carcinomas, and fully invasive carcinomas (List 5.3.4).

The observed temporal sequence of cancer development matches the predicted sequence. *In situ* (precancers) occur, on average, in the youngest observed age group, followed by microinvasive cancer, followed by invasive cancer. There is about a 15-year gap between the average age of diagnosis of *in situ* carcinoma of the cervix and fully invasive carcinoma of the cervix.

4. **Every cancer has a precancer.**
   - Corollary: If all precancers were eradicated, there would be no cancers.

If every cancer has a precancer, and if every precancer were treated, there would be no cancer. This is the fundamental argument of this book. Unfortunately, because there are thousands of different cancers in humans, and because we have not identified a precancer for all these cancers, we cannot know with certainty that our argument is correct. Nonetheless, based on our accumulated experience with precancers, we can draw a few conclusions that support our opinion. First, most common cancers of

List 5.3.3 Progressive age-at-diagnosis for breast precancers.

Median age of women with intraductal hyperplasia (IDH):	45 years (91)
Median age of women with atypical intraductal hyperplasia (AIDH):	50 years (91)
Median age of women with ductal carcinoma in situ (DCIS):	60 years (92)
Median age of women with breast cancer:	61 years (93)

List 5.3.4 Average age at diagnosis of cancers and their precedent precancers.

Average Age at Occurrence	Diagnosis
Precancers	
34 years	Carcinoma *in situ*, not otherwise specified
34 years	Squamous cell carcinoma *in situ*, not otherwise specified
35 years	Squamous cell carcinoma, large cell, nonkeratinizing, *in situ*
37 years	Squamous cell carcinoma, keratinizing, not otherwise specified, *in situ*
39 years	Adenocarcinoma *in situ*
39 years	Squamous intraepithelial neoplasia, grade III
Microinvasive Cancer	
41 years	Squamous cell carcinoma, microinvasive
Fully Invasive Cancers	
49 years	Squamous cell carcinoma, large cell, nonkeratinizing
51 years	Squamous cell carcinoma, small cell, nonkeratinizing
51 years	Squamous cell carcinoma, keratinizing, not otherwise specified
51 years	Adenocarcinoma, not otherwise specified

humans have well-described precancers (see Chapter 4). Second, in all instances where precancers have been successfully treated, cancer does not develop.

Let us examine the precancers for the five cancers that account for about 60% of cancer deaths in the United States (List 5.3.5). We have identified precancers for common tumors that account for the majority of cancer deaths. In addition, we have found

List 5.3.5 Age-adjusted U.S. death rates (per 100,000 population) for the top cancer sites (34).

Annual deaths	Precancer
Lung, bronchus: 54.1	Squamous dysplasia, AAH
Colon, rectum: 18.8	Colon adenoma
Breast: 14.1	ADH (and DIN)
Pancreas: 10.6	PanIN
Prostate: 10.1	PIN

AAH = atypical adenomatous hyperplasia of lung; ADH = atypical ductal hyperplasia of breast; **DIN = ductal intraepithelial neoplasia of breast**; **PanIN = pancreatic intraepithelial neoplasia**; PIN = prostatic intraepithelial neoplasia.

precancers for dozens of other cancers. Whether we can identify a precancer for every single cancer, from among the thousands of recognized cancers, is a problem that may have very little practical importance.

We can certainly conclude that common tumors of humans all have identifiable precancers. Can we also assume that the precancer is an obligate precursor to the cancer? Might there be a subset of cancers that do not arise from an identifiable precancer? As an example, we know that adenomas are a **precancerous condition** that may lead to the development of colon cancer. Is it true that every colon cancer is preceded by a colon adenoma? Are there cases of colon cancer arising ***ab initio*** from a single malignant cell that appeared within a population of normal cells and were not associated with an identifiable precancerous condition?

Virally induced cancers may arise very soon after **immunosuppression** of humans, sometimes in just a few months (94). When the induction time for **virally induced** tumors is short, it may be impossible to identify a precancerous lesion. The concept of precancer loses its significance when there is virtually no time between the causative event (viral replication in an immunosuppressed host) and the appearance of cancer.

Virally induced tumors can be prevented if the viral infection is avoided. Successful vaccines against hepatitis B and C and **human papillomavirus (HPV)** are expected to greatly reduce the incidence of **liver cancer** and cervical cancer. One might hope for similar results for other human viruses that cause cancer (Epstein-Barr virus, herpesvirus 6, HIV, human T-cell leukemia virus). Although we know very little about tumors that do not arise from morphologically identifiable precancers, there is reason to hope that this group of tumors have a viral causation and can be prevented with vaccines or virally effective drug therapy.

5. Treating a precancer is easier than treating a cancer.

Cancer treatment is dangerous, not just because cancer is a dangerous disease, but because the drugs and protocols used in the treatment of cancer are all capable of producing morbidity and mortality. As we have discussed previously, cancer is a lesion that is designed for relentless growth. Because cancers are genetically unstable, new subpopulations of cancer cells create a heterogeneous tumor population that is difficult to eradicate with any single medication.

Precancers are designed for regression. Most early precancers spontaneously regress. It seems obvious that it is easier to kill death-prone lesions (such as precancers) than immortalized lesions (such as cancers). Such has been the case. Many drugs that have proved effective against precancers have been remarkably safe remedies. For example, calcium has been shown to reduce the number of colon adenomas (95). A reduction of colon adenomas has also been achieved with **Sulindac**, a nonsteroidal antiinflammatory drug (96). We discuss precancer treatment in more depth in Chapter 7.

6. **A single precancer may develop into one of several closely related cancers.**
   - Corollary: There are more types of cancers than there are types of precancers.

There are thousands of different classified types of cancer (97). The number of recognized types of precancers is much smaller (65). For a given tissue, the same observation is true. A small number of recognized precancers seems to account for a larger number of recognized cancer types. Several common cancers (and several additional rare cancers) arise directly from the bronchus, so-called bronchogenic carcinomas: squamous cell carcinoma, **small cell carcinoma** (a neuroendocrine tumor), and adenocarcinoma. In addition, listings of types of bronchogenic cancer usually include large cell carcinoma, a poorly differentiated epithelial tumor corresponding to no recognized differentiated cell type. There is only one known bronchogenic precancer, squamous metaplasia with dysplasia. Careful studies of bronchial dysplasias indicate that what appears to be a monomorphic population of atypical squamous cells is actually a mixture of cells with squamous, glandular, and neuroendocrine morphologies (98). Likewise bronchogenic cancers, although they may appear monomorphic under the light microscope, typically have tripartite differentiation (squamous, glandular, neuroendocrine) on **electron microscopic** examination. This would indicate that a single precancer may develop into any of several different cancers (Figure 5.2).

How one cell type is selected to dominate in the resulting tumor is unknown. It may be as simple as luck, with one dysplastic subclone or another selected over time. Whatever the mechanism, the number of recognized precancers is far less than the number of recognized cancers. At this time, it seems reasonable to conclude that there is a one-to-many relationship between types of precancer and types of cancer.

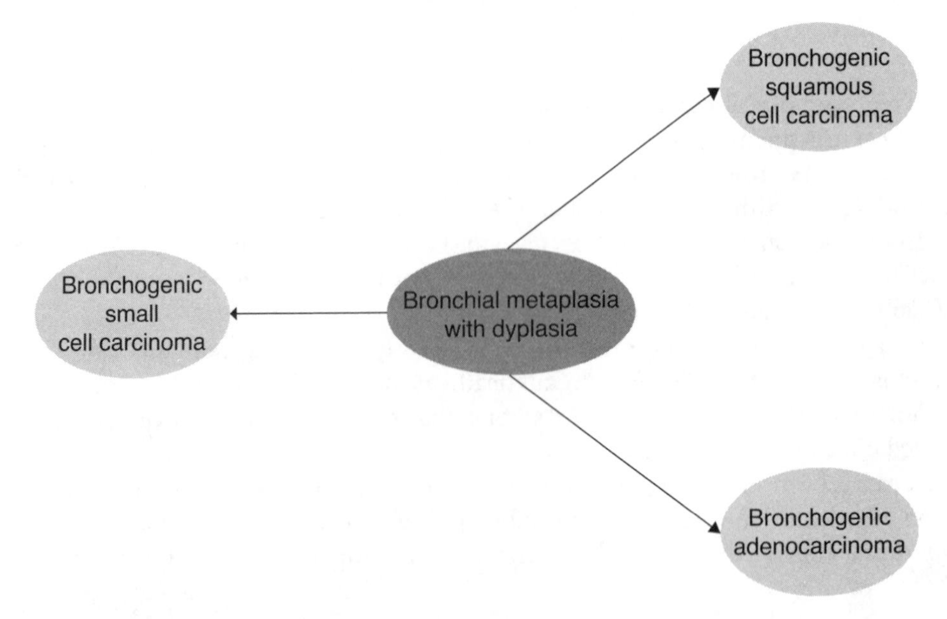

FIGURE 5.2 Squamous dysplasia of bronchial mucosa can develop into any type of bronchogenic carcinoma.

7. **Precancers contain a characteristic genetic abnormality that distinguishes one class of cancer from another class of cancer.**
   - Corollary: Agents that target an essential pathway in a particular cancer will also target the same pathway in the precancer for its corresponding cancer and for all closely related cancers.
   - Corollary: Precancers lack uncharacteristic genetic abnormalities that are accumulated by cancers during tumor progression.

As a general rule, precancers accumulate genetic and epigenetic abnormalities as they progress from mild dysplasia, moderate dysplasia, severe dysplasia, and carcinoma *in situ* (58–61). When a precancer crosses the threshold to cancer by invading normal tissue, the accumulation of abnormalities continues until the cancer is cured or until the patient dies. The phenomenon of increasing genetic and epigenetic abnormalities, coupled with increases in the degree of nuclear atypia, subclonal **heterogeneity**, and aggressive behavior, is known as tumor progression.

Because advanced cancers have many changes that distinguish them from normal cells, finding any characteristic genetic marker for a cancer is very difficult. In addition, a cancer cell within a tumor may harbor a set of abnormalities that are not present in cancer cells located elsewhere in the same tumor. These unshared abnormalities may contribute to the phenotype of a particular cell, but they do not characterize the entire tumor population. Precancers have fewer alterations than cancers, and the earliest precancers have fewer genetic alterations than later-occurring precancers. Most importantly, the earliest precancers contain a minimal set of abnormalities that characterizes the precancer and the cancers into which they develop.

The myeloid neoplasms provide the best-studied examples of simple precancer markers persisting in the cancers that later develop. There are two classes of precancer lesions for the myeloid lineage: **myelodysplasias** and myeloproliferative disorders.

Myelodysplasia (previously known as **preleukemia**) is characterized by dyserythropoiesis (inability to produce normal blood cells), pancytopenia (reduction in more than one type of circulating blood cell), platelet disorders (increased, decreased, or morphologic or functional abnormalities of platelets), and a tendency to progress to leukemia. Patients with myelodysplasia may die from **anemia** and other consequences of their blood disorder, without ever progressing to leukemia. Myelodysplasia is an example of a precancer that can cause death.

No chromosomal abnormalities are absolutely specific for myelodysplasias, but several **cytogenetic** alterations account for the majority of cases: deletion of the long arm of chromosome 5 (5q-) is present in up to 30% of cases, trisomy 8 is present in 19% of cases, and 7q- or monosomy 7 is present in 15% of cases. Myelodysplasia occasionally follows treatment with leukemogenic agents. Monosomy 7, 7q-, monosomy 5, and 5q- are the most common changes related to alkylating agent exposure. Alterations at 11q23 are generally present following treatment with topoisomerase inhibitors (99). In cases where a myelodysplasia contains a cytogenetic marker, the same marker, and numerous additional cytogenetic abnormalities, are present in acute leukemias that occasionally develop in patients with myelodysplasia.

Myeloproliferative disorders are neoplastic or hyperplastic conditions characterized by sustained growth of a stem cell subpopulation of myeloid origin that produces an increase in one or more types of circulating blood cells (such as platelets, red blood cells, or granulocytes). Myeloproliferative disorders include chronic eosinophilic leukemia, chronic myelocytic leukemia, chronic myelomonocytic leukemia, chronic neutrophilic leukemia, essential thrombocythemia, hypereosinophilic syndrome, juvenile myelomonocytic leukemia, polycythemia vera, systemic **mastocytosis**, and primary myelofibrosis. Primary myelofibrosis is characterized by sustained hyperplasia of extramedullary hematopoietic cells (myeloid growth outside the bone marrow). The **pathogenesis** of myelofibrosis is obscure and somewhat controversial. Extramedullary hematopoiesis probably results from a primary fibrosing disease of the bone marrow.

Most myeloproliferative disorders have characteristic mutations: BCR/ABL in chronic myelogenous leukemia; JAK2 mutations in polycythemia vera (100), essential thrombocythemia, and primary myelofibrosis; c-KIT mutation in systemic mastocytosis; rearrangements of PDGFRB in chronic eosinophilic leukemia and chronic myelomonocytic leukemia; and RAS/PTPN11/NF1 mutations in juvenile myelomonocytic leukemia (101). A specific JAK2 mutation is present in more than half the patients with myelofibrosis (102). JAK2 mutations can be found in blood from about 10% of the healthy population, consistent with the general observation that the earliest precancer changes are common (103). Leukemias that develop from a myeloproliferative precancer carry the same genetic abnormality that characterized the precancer and additional genetic changes that arise during tumor progression.

Chronic myelogenous leukemia (CML) has a biology that overlaps the closely related concepts of myeloproliferative disorders and leukemias. CML is a clonal disorder with an increase in the number of circulating neutrophils. Neutrophils in CML are characterized by a specific cytogenetic aberration, the **Philadelphia chromosome**, in which parts of chromosomes 9 and 22 translocate and form a fusion gene composed of part of the breakpoint cluster region (*BCR*) gene and part of the Abelson leukemia (*ABL*) gene (Figure 5.3).

The BCR/ABL mutation occurs in a hematopoietic stem cell that would normally produce any and all the committed stem cells of the various blood cell lineages (including erythroid, granulocytic, megakaryocytic, lymphocytic, and monocytic cells) and results in a massive expansion of myeloid cells (104). The expanded neutrophil population in CML are postmitotic (i.e., nondividing) cells that behave very much like normal neutrophils. Patients with advanced CML may have no physical symptoms other than abdominal swelling and splenomegaly (105). Unlike most neoplastic cells, the neutrophils of CML have virtually no nuclear atypia.

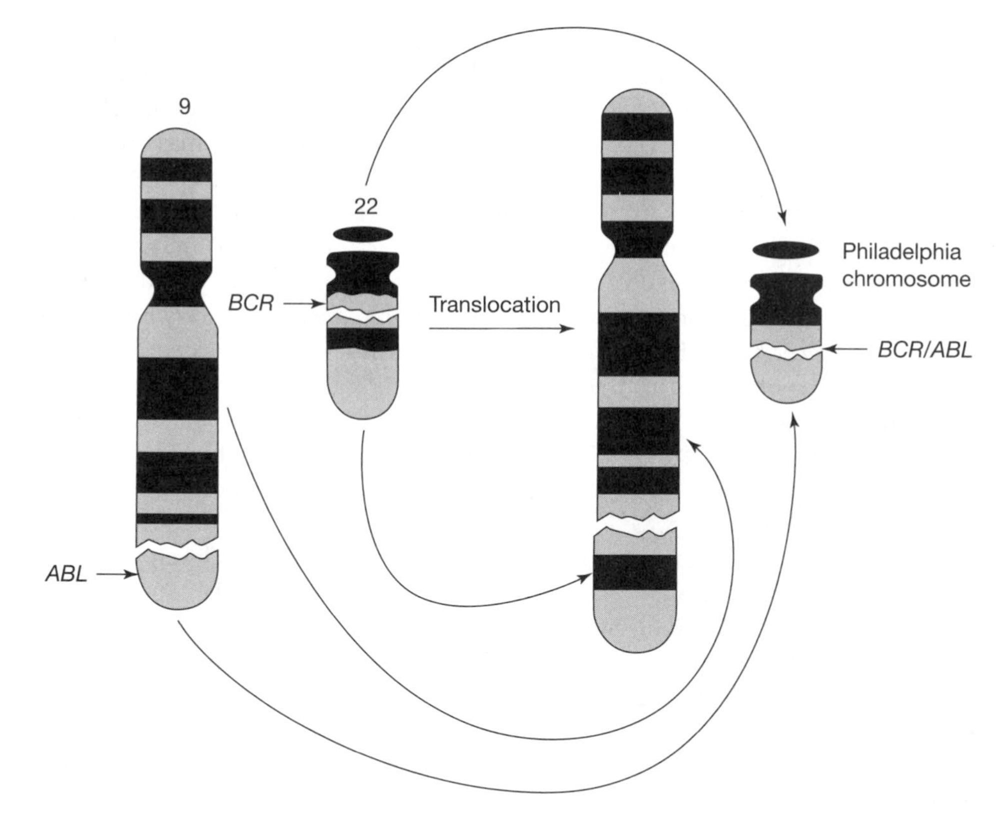

FIGURE 5.3 The Philadelphia chromosome. Reciprocal translocation between chromosome 9 and 22 results in creation of the *BCR/ABL* fusion **oncogene**. From *An Introduction to Human Disease*, Seventh edition. Illustration courtesy of Leonard V. Crowley, M.D. Century College.

After an indeterminate period of time, which may extend over years, people with CML enter an aggressive phase of disease, called blast crisis, in which immature, blastic myeloid cells expand the bone marrow and begin to appear in peripheral blood. During this phase, multiple cytogenetic changes accumulate in circulating neutrophils, and heterogeneous populations of circulating cells are seen (e.g., lymphoid blasts) (106). The development of blast crisis worsens the prognosis for patients with CML.

The biology of CML is much more like that of other myeloproliferative disorders (e.g., polycythemia, essential thrombocytosis, and hypereosinophilic syndrome) than of malignant neoplasms. The major difference between CML and other myeloproliferative diseases is the higher rate of conversion to a malignant (blast crisis) leukemia. This higher conversion rate may result from the distinctive protein product of the CML *BCR/ABL* fusion gene, a tyrosine-specific protein kinase that promotes genetic instability and accumulation of additional mutations (107, 108).

A similar story holds for the epithelial cancers. Germline *PTEN* mutations cause Cowden syndrome and Bannayan-Riley-Ruvalcaba syndrome, two inherited disorders associated with a high rate of endometrial carcinomas. *PTEN* mutations are found in 93% of sporadically occurring (i.e., noninherited) endometrial carcinomas (38). *PTEN* mutations are also found in the earliest observed endometrial precancers (38).

As precancers progress toward *in situ* cancers and invasive cancers, they accumulate additional genetic abnormalities. Some early precancers may harbor multiple genetic aberrations, and we would expect this to be the case in cancers whose pathogenesis starts with the acquisition of genetic instability, as in the case of hereditary nonpolyposis colorectal cancer (109). However, genetic abnormalities do not reverse themselves over time, and reports confirm that the earlier the precancer, the fewer the genetic abnormalities (58–62). Observations on the myeloproliferative disorders teach us that some early precancers contain pristine, characteristic alterations that yield a proliferative phenotype that persists through the progressive phases of precancer development and the emergence of cancers.

Knowing this, the precancers offer the best opportunity to develop targeted approaches against molecular alterations that drive carcinogenesis. The same opportunity to detect and target key molecular pathways in cancers is not possible for advanced cancers, because innumerable secondary genetic alterations obscure the original defect. An advanced cancer may acquire genetic aberrations that free it from dependence on the characteristic abnormalities that originally accounted for its development.

Currently, our knowledge of characteristic molecular markers for precancers is limited to just a few examples. On the basis of what we have observed, it seems that whenever we find a characteristic genetic abnormality in a precancer, the same abnormality has persisted in examples of the developed cancer. Conversely, when there is a genetic abnormality characteristic for a cancer, the same abnormality has always been present in the corresponding precancer.

8. Precancers can be distinguished from the cancers into which they develop.

- Corollary: Precancers can be diagnosed and studied.

Not every cancer researcher believes that precancers exist as lesions separable from cancers. If you think of cancer as a developmental process that proceeds through many steps, precancers become just another stage in the life of a cancer: unworthy of distinction as a separate biologic entity.

Truly, precancer is a stage in the development of a cancer. Cancers themselves continue to develop over time, accumulating genetic alterations, and becoming increasingly heterogeneous as a result of the emergence of new cancer subclones, a process called cancer progression. Nonetheless, precancers can always be distinguished from cancers. They have their own histologic morphology, different from the cancers into which they develop. They often regress, a phenomenon not present in cancers. As discussed in an earlier section of this chapter, more than one type of cancer may emerge from a single precancer. These specific properties of precancers, absent in cancers, may one day lead to new, effective methods of cancer eradication.

9. Different precancers can be divided into classes with shared biological properties. The best medical discoveries are generalizable. For example, if antibiotics were only effective against a single bacterial species, then Sir Alexander Fleming's chance discovery would have had limited medical value. Bacteria have common biological properties (e.g., cell wall, small size, etc.) that make them different from flowers, insects, and people. Knowledge of the general properties of bacteria can suggest therapeutic strategies that extend their value to all members of its class. Bacteriologists learn the names of all the different bacteria and group the different bacteria based on shared properties. Bacteriologists use their knowledge of bacterial classes to develop new antibiotics and new therapeutic strategies.

Classification efforts typically begin by listing all members of the classification domain (i.e., creating a taxonomy). Until now, there has been no effort to list precancers or to associate precancer terms with their synonyms. A given precancer may have been studied by different researchers using different terms for the same lesion. The absence of a comprehensive precancer terminology severely limits the clinical value of research that includes precancer specimens.

Until recently, there has been no effort to group precancers by shared clinical, morphologic, or biomolecular features. If an agent were discovered that induced regression of a particular precancer, there would be no organized precancer classification prompting anyone to select biologically related lesions likely to respond to the same agent.

Precancers should have a biological classification. Database annotations using this precancer classification will provide a mechanism whereby each precancer, its related precancerous lesions, and cancers known to develop from these lesions can be linked with relevant data contained in biological data sets (e.g., **gene expression arrays, proteomic arrays, tissue microarrays, pathology data sets**). A draft precancer classification has been published that contains 4700 precancer terms with 568 distinct precancer concepts (65) (List 5.3.6).

Modern classifications serve as informatics resources capable of linking, integrating, and retrieving information contained in diverse biological datasets (110). The draft precancer classification used a novel approach to disease classification, annotating precancer

List 5.3.6 Categories of precancers (modified from 65).

a. Acquired microscopic precancers
b. Acquired large lesions with microscopic atypia
c. Precursor lesions occurring with inherited hyperplastic syndromes that progress to cancer
d. Precancerous embryonic remnants
e. Acquired diffuse hyperplasias and diffuse metaplasias

terms and classes with metadata (i.e., data that describe data; Greek: *meta* = after; Latin: *data* = things given). Metadata annotations are a critical part of the data, because the annotations link data from different databases, aiding in the discovery of new knowledge relevant to precancers. The general precancer classes are:

a. Acquired microscopic precancers

These are the lesions that most people think of when they hear the term *precancer*. All the so-called intraepithelial neoplasias fall into this category. Most examples of microscopic precancers occur commonly (actinic keratosis, cervical dysplasia). They tend to be multifocal. They tend to be noninherited lesions, often with an identifiable causation (e.g., sunlight, human papillomavirus infection). They seldom occur in children. Exceptions are inherited diseases that heighten sensitivity to a causal agent, such as the early appearance of actinic keratoses in children with **xeroderma pigmentosum (XP)**. Morphologically, they tend to have a high degree of nuclear atypia. Microscopic epithelial precancers grow by subtle replacement of normal mucosa, without producing a mass, despite many replicative cycles of growth. They progress to invasive cancer while still relatively small. The term *dysplasia* is often applied to these lesions. Dysplasia, in the context of precancer, is somatically inherited nuclear atypia. Acquired microscopic precancers often have an identifiable nondysplastic stage that precedes the appearance of nuclear atypia (e.g., squamous metaplasia of bronchus, Barrett esophagus without atypia, junctional nevus, intestinal metaplasia of stomach).

b. Acquired large lesions with morphologic atypia

These lesions tend to have a uniform appearance throughout most of their long existence, even from the smallest size (i.e., they have a long, stable growth phase). When they become malignant, there is usually a morphologically apparent focus from within the large lesion that has a crowded, irregular growth pattern and marked cellular atypia that is strikingly different from the surrounding cells. This focus enlarges, becomes invasive, and is the presumed origin of cancer that develops from the precancer. These lesions tend not to regress spontaneously. They tend to be long-lived and do not progress to cancer without first growing to a large size. These lesions are often multiple but do not occur in large numbers (hundreds) unless there is a germline mutation. The prototypic acquired large precancer is colon adenoma.

c. Precursor lesions occurring with inherited hyperplastic syndromes that often progress to cancer

As far as we can tell, cancers that develop as part of an inherited syndrome must pass through a precancerous phase. The occurrence of recognizable precancers resulting from a germline cancer-causing mutation, informs us that germline mutations cannot produce cancers *ab initio* (from the beginning) but must progress through a biological phase in which additional properties are acquired, leading to the malignant phenotype. We can also infer that the germline mutations that eventually lead to cancer must be present in the precursor lesions.

Inherited precursor lesions tend to occur very rarely in the general population but may occur with high probability (sometimes 100%) in patients carrying the germline mutation. The prototypic syndromes are *RET* gene disorders. Mutations in the *RET* gene are associated with the disorders **multiple endocrine neoplasia** type IIA (MEN2A), multiple endocrine neoplasia type IIB (MEN2B), and hereditary medullary thyroid carcinoma. Lesions in this general category tend to have a single gene mutation that may be the only lesion found in the precursor lesions. Precursor lesions tend to have the morphology of simple hyperplasia, without much nuclear atypia. Precursor lesions tend to be multiple, sometimes occurring in the hundreds, and are bilateral in paired organs. These lesions tend to occur in a much younger population than the acquired precancers. The resulting cancers can also occur at a young age.

Germline mutations associated with higher-than-normal incidences of cancer are known as inherited cancer syndromes. A psychologically and clinically more useful term is *inherited precancer syndromes.* Precancers precede cancers, and if we diagnosed inherited precancer syndromes early in life, we might have time to treat the precancers and avoid the cancers. In the case of inherited precancer syndromes, a treatment that is successful against one syndromic precancer lesion will likely be effective against all of the syndromic precancers in the affected patient (as all of the syndromic precancers have the same genetic mutation). Furthermore, because the vast majority of patients with an inherited germline disease have a mutation in the same gene, an effective precancer treatment in one individual with the disorder should work for other people with the same disorder.

d. Precancerous embryonic remnants

Primitive neoplasms that occur in infancy and childhood arise through precursors that seem to be quite different from the precancer pathway observed in epithelial tumors in adults. There are no known precancers in children that are characterized by increasing grades of dysplasia leading to invasive cancer. There is, however, abnormally persistent embryonal tissue that precedes the development of childhood cancers.

The best-studied example is the **nephrogenic rest**, a type of **embryonic rest** (111). Nephrogenic rests are foci of persistent immature cells in the infant kidney. They are present in 1% of neonatal autopsies. Nephrogenic rests are also present in about one-third of Wilms tumors, a primitive tumor of childhood arising in the kidney, also known as **nephroblastoma**. In a small study, Park and coworkers found that when Wilms tumors are associated with nephrogenic rests, the rests may contain the same somatic mutation that characterizes the tumor (112). This finding suggests that nephrogenic rests are the noncancerous precursor lesions for Wilms tumor.

A similar story applies to **neuroblastoma**, a primitive neural crest tumor seen primarily in children. Beckwith and colleagues have described persistence of primitive neural crest cells in adrenals of some infants (113). These lesions are the putative precursors of neuroblastoma. Like precancers in adults, these precursor lesions are more apt to regress than progress. The incidence of radiographically detected neuroblastoma precursor lesions and small neuroblastomas (sometimes called neuroblastoma *in situ*) is much higher than the incidence of malignant neuroblastomas (85, 114).

e. Acquired diffuse hyperplasias and diffuse metaplasias

With few exceptions, acquired small focal metaplasias and hyperplasias have a very low chance of progression to cancer and have been excluded from the classification schema because they rarely result in cancer without first progressing into diffuse lesions or acquiring nuclear atypia.

Diffuse metaplastic lesions commonly precede cancers. Presumably all bronchogenic squamous dysplasia arises from squamous metaplasia. The normal bronchus simply does not have any squamous cells. The squamous cells in bronchial squamous dysplasia must have originated from a metaplastic focus or else from nonsquamous bronchial cells that differentiated directly into a dysplastic squamous phenotype.

The prototypic lesions are diffuse Barrett esophagus, diffuse intestinal metaplasia of stomach and diffuse endometrial **hyperplasia**. These lesions tend to have chronic identifiable causes (e.g., gastroesophageal reflux disease, post–lye-ingestion esophagus, chronic gastritis, long-term **tamoxifen** therapy) and tend not to regress as long as the causation persists. Small foci of dysplastic precancers may arise from the diffuse hyperplasias and metaplasias.

This class of precursors may include so-called regressing cancers, such as ***Helicobacter***-associated MALTomas and *AIDS*-associated Kaposi **sarcoma**, that can grow as multiple tumors, all of which quickly regress when the causative agent is withdrawn (e.g., after antibiotic treatment for *Helicobacter* or after normal **immune status** is restored after withdrawal of **cyclosporine** in transplant recipients). This class may also include secondary **aplastic anemia** (e.g., **benzene** toxicity), where the **bone marrow** is repopulated by an emerging population of hyperplastic cells that carry a heightened risk of progressing to acute leukemia.

10. Precancers that progress, progress into cancers and into nothing else.

Precancers have four biological options: they can regress (vanish), stabilize (stay as they are for a protracted period), grow (get bigger without invading surrounding tissue), or progress into cancers. Some of the confusion about the precancers comes from their occasional tendency to grow into large lesions. For example, colon adenomas can become quite large. Colon adenomas with a diameter exceeding 5 cm, the same size as many colon adenocarcinomas, are not uncommon.

11. All agents that cause cancers also cause precancers.

This assertion follows from the fundamental observation that cancers are preceded by precancers. Therefore, if an agent causes a cancer, it must also have caused the lesion that precedes the cancer. Because precancers occur as multiple lesions that can occur relatively quickly, we can screen putative carcinogens by testing their ability to induce precancers. This is one of the most important properties of precancers, and it is discussed in detail in Chapter 6.

12. **Cancer that develops from precancer must be a precancer clonal descendant.**
    - Corollary: Cancers arise from precancers; often the precancer is seen in the same location where the cancer arises.
    - Corollary: Cancers that arise from precancers often have characteristic morphologic features present in the precancers (74).
    - Corollary: Cancers carry the genetic alterations of their precancers.

Precancer does not turn into cancer through sudden transformation of all cells composing the precancer. There is simply no biological mechanism whereby this could occur. Progression of a precancer into an invasive cancer must occur through the emergence of an aggressive subclone from the population of precancer cells. Therefore, as we watch precancers develop over time, we might expect to see the appearance of a small subset of cells that behave more like cancer than like precancer. As time continues, the aggressive subset invades (a property not present in precancers) and grows. Eventually, the cancerous outgrowth replaces the original precancer from which it arose.

If we examine many precancers at different stages of development (i.e., different sizes, ages, and morphologies), we might expect to see some precancers closely adjacent to cancers. In fact, this is what we see. Pathologists must inspect precancers very carefully to rule out coexistence of a small focus of invasive cancer.

Chapter 5 Summary

Carcinogenesis is the process followed by progenitor cells of cancer, that leads to the occurrence of a lesion with all the biological properties required for persistent growth, invasion, and metastasis. The earliest event in carcinogenesis is **initiation**, the alteration of a cell that permits subsequent steps of carcinogensis to proceed. Following initiation is a period of **latency**, in which events proceed without demonstrable changes in **cell morphology**. During latency, there is an expansion of the population of initiated but morphologically unaltered cells. Sometime later (days, weeks, years, or decades) a morphologically altered population of cells appears, marking the end of the latency period. In experimental systems, the latency period can be drastically shortened with promoting agents that induce the appearance of many subclonal populations of growing cells that are distinguishable from normal cells. In experimental systems, most promoted subclones of initiated cells regress without forming cancers. The altered population of cells that emerges after this latency period is called a precancer.

Precancers lack the ability to invade or metastasize, and most precancers lack the property of persistent growth; therefore, most precancers regress without ever leading to cancer. A subset of precancers continues to grow, and subclones of precancer cells emerge. Eventually (days, weeks, years, decades later), a population emerges with the properties of persistent growth and invasion. When a precancer progresses, cancer is the inevitable outcome (i.e., precancers never progress into types of lesions other than cancer).

By studying precancer biology, we may be able to develop therapies that eliminate precancers and, hence, eradicate cancer.

CHAPTER

Studying Precancers to Understand Cancers

6

Order and simplification are the first steps toward the mastery of a subject.

—Thomas Mann

6.1 Background

Despite decades of intensive study, many of the most fundamental questions about cancer biology are still unanswered. Why? In a cancer cell, everything is different from normal cells, and everything a scientist might measure is changed. Chromosomes, chromatin, genes, gene regulators, proteins, RNA, mitochondria, cell membranes, **metabolic pathways**, **cell size**, cell shape, and cell morphology are all profoundly altered in cancer cells. When everything you measure is changed, it becomes difficult to separate the important changes from the less important or secondary changes. Perhaps we would be better off studying precancers. Let us look at some of the unsolved questions in precancer research (List 6.1.1).

List 6.1.1 Unanswered questions about precancers.

What are the key cellular events in precancer development and progression?

What accounts for the biological diversity of precancers?

What is the essential marker that characterizes a precancer and distinguishes one precancer from every other precancer?

Is the new generation of molecular targeted drugs likely to cure the precancers?

If you reread this list and substitute the word *cancer* for the word *precancer*, you would find that the same questions apply. This is because the most fundamental questions about cancer relate to its beginnings. If we want to learn about the earliest events in cancer development, we must study the precancers.

The purpose of this chapter is to explain the most important areas in precancer research and how they apply to the treatment of precancers, to be discussed in Chapter 7.

6.2 Classification of Precancers

Although many precancers are well known to pathologists, there are some cancers whose precancers are a subject of controversy. The best-known precancers are the epithelial precancers, accounting for the bulk of the cancers occurring in humans. Less common cancers, the cancers that derive from lymphocytes, connective tissue, muscle, bone, and vessels, have no generally acknowledged precancers (72).

There are many examples in which an indolent, proliferative lesion may suddenly transform into a much more aggressive malignancy (List 6.2.1).

An example of a myeloid lesion that progresses to cancer is myelodysplasia. Myelodysplasias are a group of proliferative lesions of bone marrow. They are characterized by marrow **hypercellularity** (too many cells in the marrow), **dysmyelopoiesis** (dysfunctional growth, resulting in an abnormal population of bone marrow cells with a shortened cellular life-span), and pancytopenia (reduction in the different types of circulating blood cells) (69). There are several types of myelodysplastic syndromes, each distinguishable by the number of **blast cells** (immature precursor cells) in the marrow and in the circulation or by the presence of distinctive abnormal blood cell morphologies. Myelodysplasias sometimes regress, particularly those with the fewest **cytogenetic** abnormalities. They can cause death from intractable anemia. Occasionally, a myelodysplasia transforms into an acute myeloid leukemia (69). Some hematologists would say that these lesions are a type of anemia, because the medical problems that result when blood counts are low usually dominate the clinical presentation. Others might suggest that these lesions are preleukemias, because they have the potential of transforming into leukemias. The myelodysplastic syndromes fit the definition of precancer provided in this book, but there is no general consensus, among hematologists and pathologists, that these lesions should be classified as precancers. Whom do you believe when the experts disagree?

List 6.2.1 Examples of nonepithelial proliferative lesions that sometimes transform into more aggressive, morphologically atypical tumors.

Chronic myelogenous leukemia → Blast transformation
Chronic lymphocytic leukemia → Richter syndrome
Low-grade follicular lymphoma → Large cell lymphoma
Myelodysplasia → Acute leukemia
Parapsoriasis en plaque → Mycosis fungoides

Lymphomas comprise neoplasms designated by the cell of origin or by the stage of development of the cell. Examples of lymphomas are follicular center cell lymphoma, marginal zone lymphoma, and immunoblastic lymphoma. No lymphoma has a recognized, identifiable precancer. We can infer, however, that prelymphomas must exist. Rosenberg and Horning have shown that up to 25% of non-Hodgkin lymphomas regress spontaneously (90). This suggests that many clonal lymphoid growths that are commonly called lymphomas would best be regarded as prelymphomas.

Helicobacter pylori is a bacterial agent that causes acute and chronic gastritis. People with long-standing *H. pylori* have an increased likelihood of developing a particular type of stomach tumor known as a MALTom*a* (mucosa-associated lymphoid tissue lymphoma). After treatment with antibiotics to eliminate the *H. pylori* infections, MALTomas often regress (115). Precancers regress; cancers do not. The regression of MALTomas after antibiotic therapy is more consistent with the biology of a precancer than of cancer.

We see many cases in which low-grade lymphomas transform into a high-grade lesions. We also see cases in which low-grade lymphomas spontaneously disappear (regress). Yet it seems that pathologists cannot agree that these progressing and regressing neoplasms are actually precancers. There are many examples of nonmalignant lymphoproliferative lesions that are not currently included among the precancers, but should be. A consensus among pathologists is lacking (72).

One of the most important projects that must be undertaken is to develop a uniform, authoritative nomenclature for prelymphomas and other nonepithelial precancers. The NCI has previously sponsored a meeting in which the basic criteria for diagnosing precancer have been formalized (73). All that remains is to agree on a comprehensive taxonomy for precancers (which lesions to include and which lesions to exclude) and to group precancers into a useful classification (65, 116, 117).

6.3 Surrogate Markers

Despite decades of testing chemicals and physical agents for cancer-causing potential, the list of recognized human carcinogens is quite short. The U.S. National Institute of Occupational Safety and Health (NIOSH) has published a list of human carcinogens (118). There are just a few dozen listed. Most of these items are variant forms of another item on the list. For example, the entire set of all human carcinogens beginning with the letter "s" comprises just six entities, all which are variant forms of one another (List 6.3.1).

It seems that the list of human carcinogens is too short to be true. Everyone assumes that the number of human carcinogens must be very large. Approximately 100,000 chemicals are currently registered for commercial use in the United States. Natural products such as mustard seeds, fruits, and cooked meats contain thousands of unidentified complex compounds. The air we breathe and the water we drink also contain many complex compounds produced by numerous unclassified sources. Basically, we live in a universe of chemical diversity. We have collected toxicity data on only a tiny fraction of these chemicals (119,120).

The National Institute of Environmental Sciences, through its National Toxicology Program, is mandated to provide the U.S. Congress with a report on carcinogens. The full report is publicly available (121). An excerpt from the report summarizes some of the problems in determining all compounds that are carcinogenic to humans.

> *The strongest evidence for establishing a relationship between exposure to any given substance and cancer in humans comes from epidemiological studies: studies of the occurrence of a disease in a defined population and the factors that affect its occurrence (122). Epidemiological studies of human exposure and cancer are difficult (123). They must rely on natural, not experimental, human exposures and must therefore consider many factors that may affect cancer prevalence besides the exposure under study. One such factor is the latency period for cancer development. The exposure to a carcinogen often*

List 6.3.1 All human carcinogens recognized by NIOSH that begin with letter "s" (118).

Silica, crystalline cristobalite
Silica, crystalline quartz
Silica, crystalline tripoli
Silica, crystalline tridymite
Silica, fused
Soapstone, total dust silicates

occurs many years (sometimes 20 to 30 years or more) before the first sign of cancer appears. Another valuable method for identifying substances as potential human carcinogens is the long-term animal bioassay. These studies provide accurate information about dose and duration of exposure and they are less affected than epidemiology studies by possible interaction of the test substance with other chemicals or modifying factors (Huff, 1999) (124). In these studies, the substance is given to one or (usually) two species of laboratory rodents over a range of doses for nearly the animals' entire lives.

Experimental cancer research is based on the scientific assumption that substances causing cancer in animals will have similar effects in humans. It is not possible to predict with complete certainty from animal studies alone which substances will be carcinogenic in humans.

A legislative stimulus for precancer research has come from the **Food and Drug Administration** Modernization Act of 1997 (FDAMA) (125). The FDAMA contains provisions for "surrogate endpoints" in drug evaluation when the endpoints seem clinically useful. This opens the door for using precancers in several surrogate roles. Foods or drugs that cause precancers to regress are likely to reduce the incidence of cancer, whereas substances that increase the incidence of precancers are likely to increase the incidence of cancers that develop from those precancers.

Carcinogens produce precancers, which precede the development of cancers. Going to the expense of developing bioassays whose endpoint is cancer is unnecessary. Testing for agents that cause precancer and inferring that these same agents will cause cancer is easier, cheaper, and faster (126, 127).

By using precancers as a **surrogate marker** for cancers, the sisyphean task of carcinogen testing may become a little less daunting. The list of chemicals to be tested can be prioritized based on the known amount of chemical in the environment, structure-based prediction of chemical reactivity, and preliminary testing in quick, in vitro screening assays. A prioritized list of carcinogens that would be difficult to test for cancer endpoints in humans or animals, might be feasibly tested in precancer model systems. One of the promising precancer test systems is the rat colon aberrant crypt model (128). After exposure to colon carcinogens, rats develop aberrant crypts, single glands with atypia. Aberrant crypts can be detected and counted. They are considered the earliest morphologic precancer in the rat. Chemicals that increase the number of aberrant crypts are putative carcinogens. Chemicals that interfere with the induction of aberrant crypts are anticarcinogens (129). Chemicals that eliminate aberrant crypts induced by carcinogens are putative precancer therapeutic agents. Like all new model systems, results are tentative and must be interpreted with caution.

The National Toxicology Program recognizes that the strongest evidence for establishing human carcinogens comes from epidemiologic studies. Rather than using rodents, with the attendant limitation that rats and humans are not always comparable, using human precancers as a surrogate marker for carcinogenic activity may be possible. Human subjects, who regularly expose themselves to many drugs and dietary supplements in the normal course of living, might be agreeable to precancer screening. When a group of persons, all exposed to the same chemical, develop an increased incidence of the same precancer, then it can be assumed that the drug is a carcinogen. Conversely, if a group of patients has a much lower incidence of precancer development compared with the control group, then these patients might have been exposed to an anticarcinogenic drug. As an example, less time may be taken to determine whether a nonsteroidal antiinflammatory drug (NSAID) reduces the number of colon adenomas in a population screened with colonoscopy than to determine whether the same inhibitor reduces the number of colon cancer deaths.

Surrogate precancer testing on humans is one of the few instances where a carcinogen assay may actually decrease the incidence of cancer in the test subject. If a precancer is detected, the subject can be treated before the cancer develops.

6.4 Biorepositories

Considerable cancer research is conducted with samples of human cancers. The National Cancer Institute and other agencies throughout the world have spent enormous amounts of money on biorepositories that archive samples of cancers for research. *Biorepository* is a general term for any collection of biological specimens (tissues, blood, body fluids, plants, insects, seeds, and so on). Biorepositories for human tissue samples have some combination of fixed tissues and **frozen tissues**. Fixed tissues are samples that have been immersed in formalin or alcohol **fixatives** that inactivate enzymes and immobilize structural elements of cells. Once fixed, tissues can be stored indefinitely. In most instances, fixed tissues are embedded in paraffin. Paraffin-embedded, fixed tissues lose many cellular activities that may be of interest to researchers. For this reason, biorepositories often include frozen tissue samples. Freezing tissues involves obtaining a tissue specimen within a few minutes after it has been removed from the patient, freezing the tissue to a very low temperature, and keeping the tissue at that temperature indefinitely. The logistics of procuring tissues, annotating them with detailed clinical data, creating a useful database for all the tissues, holding the archived samples, disbursing them to scientists, and performing all these functions in an ethical, safe, fair manner, in compliance with federal, state, and local regulations, is a complex and expensive endeavor.

At this time, general biorepositories for precancers do not exist. Nobody has bothered to fund them. Precancers are received as routine specimens in surgical pathology departments. Actinic keratoses, dysplastic nevi, cervical intraepithelial neoplasias, Barrett esophagus, colon adenomas, myelodysplasias, and myeloproliferative disorders are lesions that every pathologist has encountered. Most precancers are small. In the past, pathologists were reluctant to donate an entire precancer biopsy to a biorepository.

Today, collecting libraries of DNA species obtained from microdissected samplings from precancer tissues is possible (130). A biorepository composed of snippets cut from small precancer biopsies suffices for many studies.

6.5 Precancers at the National Cancer Institute

There is a growing interest in precancers at the NCI, which has issued an innovative Request for Applications (RFA) for precancer research. The RFA focused on breast precancers and was released on May 30, 2007 (131). Here is an excerpt.

> *The NCI, as well as experts in the extramural scientific community recommend further research related to the biology of the pre-malignant state in human breast cancer. An expert panel convened at the November, 2004, NCI Workshop on Pre-Cancers identified delineation of the biological, genetic, and functional characteristics of pre-cancers as major scientific needs (Cancer Detect Prev. 2006;30(5):387–94). The distinctive early lesions that occur have characteristic properties that should permit them to be detected, diagnosed, and prevented from progressing to invasive cancer. The Workshop participants noted a number of impediments to conducting research on pre-cancers, including: insufficient understanding of normal and pre-cancer biology; limited access to appropriate specimens; a highly subjective, histology-based classification scheme; and the lack of strategic partnerships among research communities.*

Support for precancer research is growing at the mecca for cancer funding.

Chapter 6 Summary

The FDA's Modernization Act seems to provide some relief from the creeping fear that neither drug companies nor the NCI nor society have the resources to test all environmental chemicals for carcinogenicity or the money and patience to conduct randomized clinical trials for all promising new anticancer drugs. Precancers can now serve as surrogate endpoints for some cancer studies. Because all cancers are preceded by precancers, we do not need to waste our time and money testing for agents that induce cancers. If an agent induces a precancer, we can infer that it is a carcinogen. Likewise, agents that prevent or eliminate precancers will presumably prevent and eliminate cancers.

Modern techniques for detecting very small lesions throughout the body, and for performing genetic evaluations on very small aggregates of cells, permits the detailed evaluation of precancer lesions. These studies require access to well-preserved, well-characterized tissue samples of many different types of human precancers, held in biorepositories. To produce data that are comparable among many different laboratories, we need a classification of precancers that is used consistently by all researchers.

PART

Eradication of Cancer By Treatment of Precancers

CHAPTER

7

Treating Precancers

The superior doctor prevents sickness; the mediocre doctor attends to impending sickness; the inferior doctor treats actual sickness.

—Chinese proverb

7.1 Background

Cancer is the second leading cause of death in the United States. The lifetime risk of developing cancer is approximately 41% (5, 132). Despite the high risk of cancer occurrence, cancer prevention may not have the life-extending effect that people expect. In 1978, Tsai and coworkers calculated the increase in life expectancy that would occur if cancer was eliminated as a human disease. They predicted that the elimination of cancer would extend human life by no more than 2.5 years (133). A more modest reduction in cancer deaths (30% reduction) would increase life expectancy by 0.71 years. A similar 30% reduction in motor vehicle crash deaths would extend life expectancy by 0.21 years (133). Given the difficulty we have had in reducing the cancer death rate by even a few percentage points since 1969 (14), a cynic might think that our efforts might have been better spent teaching U.S. citizens how to drive safely.

Why is the benefit so minor? Cancer is a disease that occurs predominantly in the geriatric population. Elderly people, who do not die of cancer, die from some other disease. Eliminating one disease of the elderly allows other diseases of the elderly to take its place. The net effect of eliminating a single disease from the geriatric population is minor.

The United States pays more for medical research and more for medical care, on a per capita basis, than any other country. How has this improved the life expectancy of American citizens? In 2008, the United States ranked 47th in life expectancy among the countries of the world (List 7.1.1). The United States ranks well below countries that spend relatively little on medical research. In fact, a dozen countries and large municipalities have life expectancies that are more than 2.5 years longer than that of Americans: Andorra, Macau, Japan, Singapore, San Marino, Hong Kong, Canada, France, Sweden, Switzerland, Australia, and Guernsey. This means that there are many millions of people who have a life expectancy today that exceeds any increased life expectancy that Americans can expect to enjoy after cancer is eradicated as a disease.

Why are the citizens of these other places living longer than Americans, who live in a country with unequaled biomedical resources? The greatest improvements in the physical well-being of a population come from simple public health measures, not from complex, technologic advances.

In Finland in about 1970, the population had a shockingly high incidence of cardiovascular disease. Public health measures were introduced to improve diet, exercise, and other health habits (particularly smoking) in the population. The multidecade effort, called the North Karelia project, stands as one of the greatest successes in public health intervention (List 7.1.2). The lung cancer death rate dropped 70%. These findings emphasize that an approach aimed at improving the general health of populations, including measures that reduce the incidence of cancer, may yield results that vastly exceed the benefit that would follow from advances in cancer treatment.

The purpose of this chapter is to explore the simple, readily available measures that can accomplish these goals (List 7.1.3).

List 7.1.1 CIA World Factbook 2008, Rank-order life expectancy at birth, 2008 estimates, in years, last updated July 15, 2008 (134).

Rank	Country	Rank	Country
1	Andorra: 83.53	40	Jordan: 78.71
3	Japan: 82.07	41	Puerto Rico: 78.58
18	Italy: 80.07	43	Bosnia-Herzegovina: 78.33
22	Spain: 79.92	46	Cyprus: 78.15
25	Greece: 79.52	47	United States: 78.14
37	United Kingdom: 78.85		

List 7.1.2 Results of the North Karelia project (135).

Increased life expectancy (7 years longer for men, 6 years longer for women)
Lung cancer death rate reduced by 70% in North Karelia
Heart disease mortality reduced by 65% in men

List 7.1.3 Precancer interventions requiring small measures and no scientific breakthroughs.

1. Reduce the incidence of precancers
2. Treat precancers
   - **Early detection** followed by excision, as has proved highly effective in cervical intraepithelial neoplasia
   - Vaccines, such as Gardasil
3. Encourage the regression of precancers
4. Encourage the persistence of precancers
5. Discourage the progression of precancers

7.2 Past Success of Precancer Treatment

Many more people are alive today as the result of precancer treatment than are alive due to the treatment of cancers. The successful reduction in deaths from cervical cancer is a good example of the effectiveness of precancer treatment. Cervical cancer is a type of squamous cell carcinoma that develops at the junction between the ectocervix (the squamous-lined epithelium) and the endocervix (the glandular-lined epithelium) in the os of the uterine cervix of women. Before the introduction of cervical precancer treatment, cervical carcinoma was one of the leading causes of cancer deaths in women. Today, in many countries that have not deployed precancer treatment, cervical cancer is still the leading cause of cancer deaths (136, 137). The relatively low number of cervical cancer deaths in the United States is the result of a 70% reduction in age-adjusted mortality after the introduction of organized Pap smear screening (138–140). No effort aimed at treating invasive cancers has provided an equivalent reduction in the number of cancer deaths.

Today we know that almost all cervical cancer is the result of infection by one of several strains of human papillomavirus (HPV). The strains of human papillomavirus that cause cervical cancer are transmitted during sexual intercourse by men infected with the virus. In the late 1940s the viral cause of cervical cancer was unknown. We did know that squamous cells sampled from the uterine os had highly characteristic morphologic appearances that varied with the time of sampling (during a woman's estrous cycle) and the histologic compartment of cells (near the basal layer of the epithelium or nearer the surface layer of the epithelium). We soon learned that by sampling and examining cervical specimens from women, we could accurately determine if precancerous changes were present. If precancerous changes were present, a gynecologist could remove a superficial portion of the affected epithelium, and this would, in the vast majority of cases, stop the cancer from ever developing.

Thanks largely to the persistence of Dr. George N. Papanicolaou and his coworkers, a screening test was developed to detect cervical precancers. A typical Pap smear contains about 20,000 cervical cells, and every cell must be inspected to rule out dysplasia and other pathologic abnormalities. A large cytology laboratory can handle hundreds of thousands of Pap smears in a year.

Morphologic and epidemiologic observations on Pap smears have provided clues that have led to the identification of several strains of HPV as the major causes of cervical cancer. Recently, a vaccine against carcinogenic strains of HPV has been developed. If all goes well, most cases of cervical cancer will be prevented by this vaccine. Screening will shift away from cytology screening and toward HPV testing. The Pap smear industry will shrink as fewer and fewer women test positive for cervical precancer.

Pap smears are obtained as part of a routine gynecologic examination. They are one example of the enormous benefit obtained by including precancer detection and treatment within the scope of a common procedure intended to improve the general health of a population.

7.3 Old and New Paradigms for Treating Cancer

List 7.3.1 presents a paradigm for treating cancer.

With precancers, we may be able to skip most of these steps, advancing directly to treatment, without stopping to detect or diagnose the lesions. This is because treatment for precancers may carry little or no risks. If a precancer can be eradicated with a

List 7.3.1 Old paradigm for treating cancer.

1. Detect the cancer (usually involves recognizing a sign or symptom or suspecting cancer on a screening test).
2. Diagnose the cancer (usually involves obtaining a tissue sample through a surgical procedure and sending the sample to a pathologist, who renders a diagnosis that includes the type and grade (level of malignancy) of tumor. Diagnosis is sometimes supplemented with special studies such as cytogenetic analysis).
3. Stage the cancer (determining how widely the tumor has spread at the time of diagnosis).
4. Treat the cancer (one or more of surgery, chemotherapy, and radiation therapy).
5. Follow-up.

relatively nontoxic systemic drug, or if the transition from precancer to cancer can be delayed with hormonal manipulation, or if the initiation step of carcinogenesis (leading to precancer development) can be blocked with a dietary supplement or vaccine (e.g., **Gardasil** for cervical precancer), why not just forego the detection/diagnosis/staging steps?

The idea of receiving medical treatment for undiagnosed diseases is not new. How many people in the United States take statins, even though they have no reason to think that any of their arteries are significantly blocked by atheroma (never had stroke, never had angina, never had claudication, etc.)? How many people in the Unites States are treated for hypertension even if they've never had any of the associated diseases (never had renal failure, never had a stroke, etc.)? Virtually everyone in the United States has been vaccinated for diseases they do not have (polio, smallpox, hepatitis, etc.). A vaccination against hepatitis increases life expectancy by decreasing the incidence of hapatitis and by preventing the development of liver cancer.

Intelligent people accept treatment for diseases they do not have, because they know how bad such diseases (such as myocardial infarction, stroke, kidney failure, polio, etc.) can be. So why not start treating precancers in high-risk people who have no detected precancers? Personally, I'd rather accept treatment for a precancer, that I may not actually have, than for a cancer, whose diagnosis is certain.

7.4 Reduce the Incidence of Precancers

Carcinogenesis begins with the exposure of normal tissue to an initiating carcinogen. This first exposure creates alterations in cells (primarily mutations), that eventually lead to the emergence of precancers. The time between the initiating exposure with a carcinogen and the eventual emergence of precancers is highly variable. In the case of actinic keratoses in humans, skin exposure to ultraviolet light begins in youth. Actinic keratoses are most commonly seen in middle-aged and older individuals. A much shorter initiation-to-precancer interval has been observed with oral squamous dysplasia in tobacco chewers. The habit of chewing tobacco is not uncommon among adolescent boys. Precancers can occur while the boys are still in their teens. When an adolescent boy stops chewing tobacco, he prevents squamous precancers, thus reducing his chance of developing oral cancer. He also improves his overall health by reducing the systemic toxicity produced by chronic nicotine exposure.

The identification of initiating carcinogens and the removal of these carcinogens from the environment has been a mainstay of cancer prevention. When a carcinogen is removed from the environment, both the precancer and the cancer are prevented. Although we have managed to identify many initiating carcinogens, the key genetic and epigenetic events that define initiation, the response of cells to initiating agents, and the events that transpire between the moment of inititiation, and the emergence of precancers, are all shrouded in mystery.

As discussed in Chapter 6, early events in carcinogenesis are studied by manipulating a model biological system (typically a rodent carcinogenesis model) and counting the number of animals that developed cancer. For example, you might expose a set of rats to an initiating carcinogen. Half the rats might be treated with another compound, such as an antioxidant. If the number of rats developing cancers was smaller in the treated group (antioxidant and carcinogen) than in the control groups (carcinogen only, antioxidant only, or no treatment) , then the antioxidant might be considered an effective anti-cancer agent for the carcinogen used in the study.

There are some problems with this animal model approach. First, these studies are long and expensive. Cancers take a long time to develop, but precancers take a short time to develop. If the experimenters had looked for precancers as an endpoint, then the experiment might have been shorter and less expensive. Secondly, the experiment blurs the steps leading to the development of cancer. The antioxidant may have worked by interfering with initiation events, events occurring after initiation and before precancer emergence, modification of precancer regression, modification of precancer persistence, and/or modification of precancer advancement (to an invasive cancer).

By using cancer as the endpoint, experimenters answer the bottom-line question: "Can this agent reduce the incidence of cancer in a rodent model?" The model fails to answer a more easily answered question: "How does this agent modify the emergence of precancers?" Further studies, using precancer as the biological endpoint, might identify agents that specifically modify the regression, persistence, or advancement of precancers. A combination of drugs acting on different steps of precancer development may be more effective than anticancer drugs that influence a single, undetermined step in carcinogenesis.

The practical importance of using precancers as the measure for carcinogenicity is that studies can be performed in human subjects, yielding clinically applicable results. For example, if you suspect that a drug increases the incidence of oral cancer (based, perhaps, on preliminary epidemiologic data), you may not need to do a prospective study of cancers occurring in stratified human populations. Studying the occurrence of oral precancers (squamous dysplasias) might suffice. In general, the sudden emergence of precancers in a human subpopulation is a strong indicator of worse things to come.

7.5 Detect and Treat Precancers

Until recently, all precancer treatment involved precancer detection followed by precancer **excision**. As discussed earlier, the greatest success of detection/excision has been with cervical precancers (138–140).

There is a common misperception, held by laypersons and healthcare professionals alike, that the Pap smear is a test for cervical cancer. This is simply not the case. Although cervical cancer cells can be detected on Pap smear, by the time most cervical cancers have developed, the tumor is visible through the vaginal speculum and needs to be biopsied, staged, and treated. Advanced-stage cervical cancers have a poor prognosis. The purpose of the Pap smear is to screen an asymptomatic, general population of women to determine if they have precancerous lesions that can be easily excised, preventing the development of cancer.

Colon cancer is the second most common cause of cancer death in the United States. In 2008, 49,960 Americans died of colon or rectal cancer (2). Colonoscopy is a diagnostic tool that is used to assess many symptomatic lesions in the colon, including ulcerative colitis and other inflammatory diseases of the colon and sources of colonic bleeding. Colonoscopy is a screening tool for the general population over age 50 and for younger age groups who have an inherited or acquired increased risk for colon cancer. Almost all colon cancers develop from colon adenomas, a noninvasive precancer. By finding and excising colon adenomas during colonoscopy, the incidence of colon cancers can be dramatically lowered.

Squamous cell carcinoma of skin is the most common cancer in humans, accounting for about 600,000 new cases each year in the United States. Actinic keratosis, the precursor of squamous cell carcinoma of skin, is the most common precancer in humans. These lesions occur on sun-exposed skin, particularly arms, neck, face, and the dorsa of hands. Although squamous cell carcinoma of skin is common, actinic keratoses greatly outnumber squamous cell carcinomas. In tropical areas, multiple actinic keratoses occur on most fair-skinned adults. Agencies that collect cancer data, such as hospital tumor registries, the CDC, and the NCI SEER program, do not include squamous cell carcinoma of skin or actinic keratoses in their databases. Basically, these lesions are too numerous to bother counting.

Both actinic keratosis and squamous cell carcinoma of skin are easily detected (as blemishes) by patients, who are apt to seek treatment at early stages of disease. Actinic keratoses are treated by simple excision or, if numerous, by freezing with liquid nitrogen or by topical application of a cytotoxic agent. By treating the precancer, the cancer never develops. Melanoma is known as one of the most malignant tumors in humans, capable of metastasizing widely throughout the body, even while the primary skin **tumor** is still small and inconspicuous. Although melanoma can metastasize while the primary lesion is small, almost every malignant melanoma arises from a precancerous lesion or from a so-called "thin" cancer that has very little potential for metastasis. A trained observer can evaluate pigmented skin lesions by visual examination and can often determine which ones are benign and unlikely to have any clinical significance, which ones are precancerous and must be removed, which ones are "thin" melanomas and must be removed, and which ones are likely to be malignant melanomas requiring a wide surgical excision and careful staging to determine whether metastasis has occurred. All suspicious pigmented lesions should be biopsied for diagnosis by a pathologist.

The common nevus (also known as mole) is a small, pigmented lesion, composed of round nevus cells, derived from the same cytologic lineage as skin melanocytes, forming nests in the lower epidermis or in the dermis. The common nevus is a benign lesion. Melanomas may arise from dysplastic nevi, precancerous lesions that must be distinguished, by pathologists, from the common nevus. Cells in a dysplastic nevus show nuclear atypia. The nests of dysplastic nevus cells are accompanied by architectural changes in the dermis adjacent to the epidermis.

Melanomas usually develop in a sequential manner, that begins with a so-called radial growth phase, during which malignant melanocytes grow horizontally, through the epidermis and the upper dermis, without invading deeply. During this radial growth phase, the melanoma is thin, and metastasis seldom occurs. Following the radial growth phase, melanocytes begin a vertical growth phase and invade deeply into the dermis and subcutaneous tissue. During the vertical growth phase, metastasis is likely to occur. If melanomas are diagnosed and excised while still thin, the patient has an excellent chance of achieving a complete cure. If the melanoma is excised after it has begun its vertical growth phase and has metastasized, the chances of achieving a complete cure are greatly reduced.

Visually screening for melanoma precancers and for thin melanomas is a practice that is followed by all dermatologists. The expectation is that detecting and excising dysplastic nevi and thin melanomas will reduce the death rate from melanoma. The melanoma death rate has been rising steadily over the past several decades (Figure 7.1). The U S. death rate from melanoma for white males has more than doubled since 1969. Has surveillance for dysplastic nevi reduced the melanoma death rate? Making a strong case for the effectiveness of precancer treatment when the death rate keeps rising is impossible. We simply do not know the effect of surveillance on the melanoma death rate. Nonetheless, prudent people usually opt to excise a pigmented lesion with clinical features suggestive of precancerous melanoma.

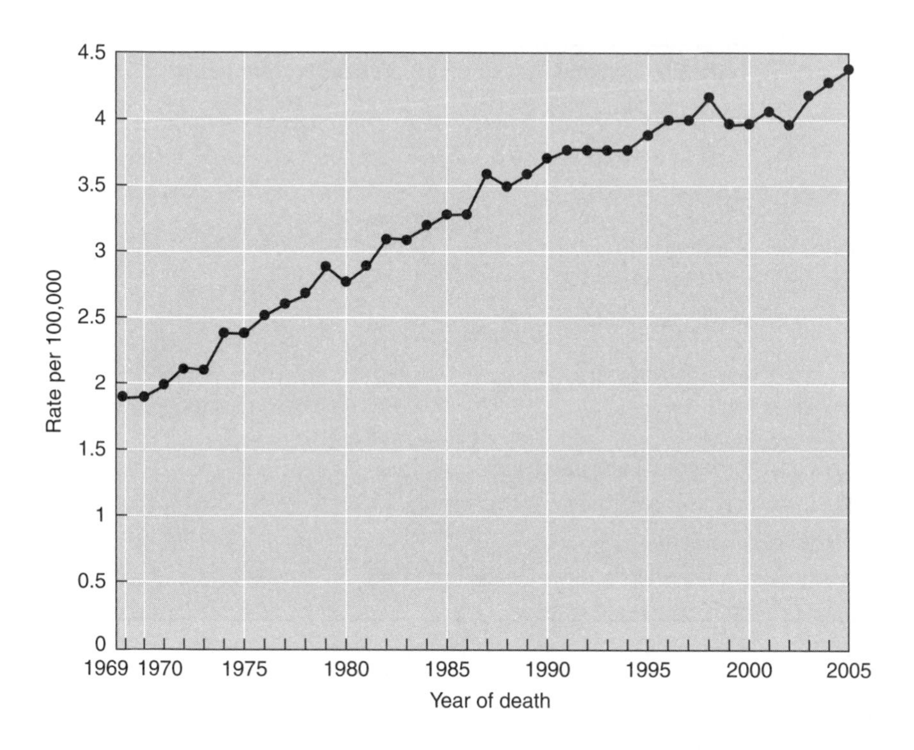

FIGURE 7.1 U.S. death rate from melanoma in white males, 1969–2005. Data provided by the NCI's SEER program (http://seer.cancer.gov/cangues/mortality.html).

7.6 Treat Precancers without Detection

Many precancers can be treated directly using surgery or medications. Currently, surgical remedies are used for cervical (cervical intraepithelial neoplasia), skin (actinic keratoses and dysplastic nevi), and colonic (adenomas) precancers. The only drawback with this approach is that the lesions must be detected before they are treated. Somehow, either by direct inspection, cytologic smear, or endoscopic examination, the lesion must be visualized. Precancers that are small and internal may be difficult to detect with current imaging techniques. Examples include pancreatic intraepithelial neoplasia, precancers of sarcomas (often located deep within muscle, connective tissue, or bone), and precancers of the central nervous system. Many medical treatments have proved effective in humans or animal models, when treatment is begun before the precancer is detected (List 7.6.1).

List 7.6.1 Medical (nonsurgical) precancer treatments.

Chemical agents
Sulindac (colon) (96, 128, 141)
Tamoxifen (breast) (142)
Hormone cessation (breast) (143)
Antioxidants (stomach) (144)
Progestin for endometrial precancer (145)
Hypomethylating agents (myelodysplasia) (146)
Dietary supplements (calcium, flax oi) (95, 129)
Vaccines (147)
 HPV vaccine for cancer of the uterine cervix
 Hepatitis B vaccine for liver cancer
Antimicrobials
H. pylori eradication, followed by MALToma regression (115)

Many more medical treatments are awaiting clinical trials. The kinds of precancers that may be most amenable to clinical trials are those whose progression can be followed over time to determine if a candidate treatment is effective (List 7.6.2).

The key point to remember is that medications that treat precancers are sometimes nontoxic. Some of the precancer medications under consideration may actually have beneficial effects on the general health of patients (such as calcium supplements, anti-inflammatory agents, antioxidants, antimicrobials). In such cases, prophylactic treatment of people at high risk for developing precancers, and even the general population may be a practical option.

List 7.6.2 Known precancers that may serve as treatment candidates and available methods to gauge treatment response.

Multifocal squamous dysplasia in oral cavity (look in mouth, biopsy)
Barrett esophagus with dysplasia (endoscopy, biopsy)
Ductal intraepithelial neoplasia of breast (mammography)
Prostatic intraepithelial neoplasia (biopsy)
Adenomas of colon (colonoscopy, biopsy)
Dysplasia of ulcerative colitis (colonoscopy, biopsy)
Atypical adenomatous hyperplasias of lung (spiral CT)
Endometrial intraepithelial neoplasia (biopsy, sonogram in some cases)

7.7 Encourage Regression of Precancers

Although precancers often regress spontaneously, we know absolutely nothing about the biological mechanism of regression. This phenomenon may be the most fertile area in cancer research, yet there is virtually no funded work in this field. There are several reasons for this, the most obvious being that at this time no distinction is drawn between agents that enhance regression and agents that kill precancers. In either case, the precancer disappears.

Most cancer chemotherapeutic agents directly kill their target cells. Through one mechanism or another, these agents exert a toxic effect by attacking organelles, cellular consituents, or pathways. We can hope that some of the agents that can kill cancers will also kill precancers.

A distinction can be made between agents that kill precancer cells and agents that enhance precancer regression. Regression is a normal phenomenon in precancers. We do not know much about how regression works, but agents that enhance regression exert an influence on normal pathways and are not likely to be toxic agents. They will most likely fall into a pharmacologic category that contains a class of molecules different from the class that kills precancer cells.

Agents that enhance precancer regression may be the best way to eliminate precancers. We will not find agents that target precancer regression unless we search for them, by studying the biological process of regression.

7.8 Encourage Persistence of Precancers

Interventions that delay the onset of cancer can be very important. Cancer is predominantly a disease of the elderly. If the average age of cancer occurrence is pushed back another 10 or 20 years, many persons in the population at greatest risk for developing cancer will have died of other causes.

7.9 Discourage Progression of Precancers

What happens when precancers do not progress? Consider this example in the article, "Reversing Trend, Big Drop Is Seen in Breast Cancer," by Gina Kolata, in the December 15, 2006, issue of *The New York Times.*

> *Rates of the most common form of breast cancer dropped a startling 15 percent from August, 2002, to December, 2003, researchers reported yesterday.... The reason, they believe, may be because during that time, millions of women abandoned hormone treatment for the symptoms of menopause after a large national study concluded that the hormones slightly increased breast cancer risk.*

There is a problem with the kind of reasoning that Ms. Kolata describes. The report on the cancer risks of hormone treatment for menopause came out in July, 2002. The drop in breast cancer rates occurred between August, 2002, to December, 2003. For most other diseases, this might lead you to suspect causality. But carcinogenesis is a multistep process that extends over many years. There is no reason to think that a decision to discontinue a drug associated with an increased cancer risk will result in an immediate (next month) drop in cancer incidence.

That breast carcinogenesis takes about 15 years (from initiating event to tumor detection) is commonly accepted. Let's assume that the 15-year number is good. Then if there is a large drop of cancers noted beginning August, 2002, wouldn't it make sense to look for some change in carcinogen exposure that occurred starting in August, 1987 (not the summer of 2002)?

The only way around this objection is to assume that women who opted out of hormone treatment in August, 2002, and who (might have) accounted for the drop in breast cancer incidence starting the following month, had a lesion that was not cancer, whose development into cancer was stopped when hormone treatment was omitted—and that lesion would be called a precancer. If you look at the observations this way, it changes the question to, "Has the reduction in hormone treatment in menopausal women resulted in a decrease in the progression of precancerous breast lesions?"

Chapter 7 Summary

The lessons learned from all these observations is that simple measures, using no special technology and requiring no miracle breakthroughs, can yield enormous health benefits. Measures that reduce the number of precancers, treat the precancers, increase the rate of precancer regression, lengthen the persistence of precancers, and halt the advancement of precancers are all practical interventions that may eradicate cancer. Moreover, when included in a general plan to enhance the overall health of patients, precancer treatments may lengthen human life expectancy.

CHAPTER

Politics of Precancers

To be blunt about it, there's no money in prevention, and once you've got cancer you'll pay anything to try to stay alive.

—Peter Montague (from his article, "The Environmental Causes of Cancer: Why We Can't Prevent Cancer")

8.1 Background

PubMed is the National Library of Medicine's bibliographic database. Anyone can search the 18 million citations in the PubMed database by visiting their website. In May, 2008, I searched PubMed for citations related to cancer and to precancer. The results are in List 8.1.1. Each term is followed by the number of PubMed citations (research articles) that have the term contained in the title.

These numbers suggest that precancers receive a miniscule fraction of the research effort devoted to cancers. In Chapter 3, we saw that that the bulk of funding in the cancer field is directed toward the most frequently occurring cancers, that are responsible for the greatest number of cancer deaths in the U.S. population (e.g., lung, colon, breast, and prostate cancers). Never mind that mortality from these common cancers has scarcely budged since the war on cancer was officially declared in 1971. Cancer researchers spend too much of their time and money trying to find a cure for advanced, common cancers; not enough time and money is spent on the precancers.

List 8.1.1 PubMed citations with cancer and precancer synonyms in their titles.

Cancer titles

1. Cancer, 388,888
2. Tumor, 143,219

Precancer titles

1. Precancer, 347
2. Premalignant, 1,314
3. Premalignancy 66
4. Intraepithelial neoplasia, 2,533
5. Preneoplastic, 1128
6. Preneoplasia, 92

All humans have a deep-rooted fear of cancer. There are very few people living today who have not lost close friends and relatives to cancer. People who have cancer are willing to pay almost anything for a cure. No wonder the National Cancer Institute receives much more funding than any other disease-based institute at the National Institutes of Health.

While cancer evokes strong, visceral emotions, precancer does not. Nobody has seen a child who has died from precancer. Nobody wakes up startled, in a cold sweat, from a precancer nightmare. Few people see the necessity of finding a cure for a disease that they do not have (or do not know that they have). Evoking a strong emotional reaction for a relatively benign disease that precedes another disease is nearly impossible.

The mission of the National Cancer Institute is to eradicate the death and suffering from cancer through scientific research. You would think that an institute motivated by science would approach the problem of cancer eradication objectively. This is not the case. All research is conducted in a environment where many different ideas compete for funding. Precancer researchers will have a hard time receiving funding if there is a preponderant opinion that research for common malignant tumors is more urgent than precancer research.

Oncologists have a very large role in planning new research initiatives at the National Cancer Institute. They are involved in selecting candidate agents, designing treatment protocols, accruing patients for trials, and treating patients in every new clinical trial coordinated through the National Cancer Institute. Oncologists are people who devote their lives to treating people with advanced cancers. All clinical trials are designed for patients with advanced cancers. Oncologists have no role in the diagnosis or treatment of precancerous lesions. Convincing oncologists to divert their attention away from cancer clinical trials and toward trials designed to treat precancers is very difficult.

Over the decades, I have had many conversations with oncologists at the National Cancer Institute trying to convince them that more money should be spent on precancer trials. I have never gotten a single oncologist to agree with me on this subject. The counterarguments that I have heard (paraphrased) fall into several broad categories: nonexistence arguments (precancers do not actually exist); irrelevance arguments (precancers fall outside the realm of cancer research); priority arguments (nobody really cares about the precancers, and just about everything else in the cancer field is more important); and impracticality arguments (treating the precancers is impractical). Here are the antiprecancer arguments with my responses.

8.2 Nonexistence Arguments

Argument 1.

There is no such thing as precancer. The lesions that are called precancers are simply early (or small) cancers.

Response. Leslie Foulds is one of the most thoughtful early pioneers in the field of carcinogensis. In 1958 Foulds wrote, "Lesions described as 'precancerous' are visible steps in a dynamic process of neoplasia; these lesions may or may not undergo progression to a more advanced stage of neoplasia." This description of precancerous lesions places precancers in a transitional state that is encountered during the multistep progression from normal cells to cancer cells. People who think of carcinogenesis as a gradual progression of genetic lesions and phenotypic properties leading to invasive cancer tend not to think in terms of distinct or even definable lesions occurring en route.

Why can't we recognize that precancers are just an early form of cancer and refer to the precancers by the name of the corresponding developed cancer (e.g., early squamous cell carcinoma, early adenocarcinoma)? Wouldn't that make life a lot easier than naming and characterizing a new disease entity for the preinvasive stage of every cancer? Much as we like data simplification, this just can't be done in the case of the precancers.

Precancers have properties that separate them from cancers. They are not simply small, or early, versions of cancers. These properties, particularly spontaneous regression and transition from noninvasion to invasion, deserve to be studied. If we study the biology of the preinvasive stage of cancers, then we need to have standard names and diagnostic criteria for this stage, so that all investigators use the same morphologic and biologic features to identify the same named lesions. Otherwise, research results will not be comparable from laboratory to laboratory, and the field of precancer research will not advance.

Argument 2.

The transition from precancer to cancer is characterized by the acquisition of invasiveness. There is no practical way, however, to determine the precise moment that invasiveness is acquired by a lesion. There is no practical method, therefore, to reliably distinguish a precancer from a cancer in every instance (i.e., there is no way to be confident that a lesion has not acquired the ability to invade). Precancers, therefore, have no validity as biological entities.

Response. This argument, esoteric as it may seem, has been discussed in the pathology literature (116, 148). Basically, biologists are not adept at determining the precise moment of naturally occurring events. We cannot determine the exact moment, during myeloid proliferation, when a promyelocyte becomes a myelocyte or the moment when an undifferentiated collection of embryonic cells becomes a fetal organ. We cannot even determine, to everyone's complete satisfaction, the moment of human death. Everyone accepts the existence (and inevitability) of these transitional states, without defining the moment of transition.

The development of precancers and cancers is punctuated by changes that we can neither quantify nor specify. The earliest precancers are the earliest recognizable lesions that precede the development of cancer. In experimental systems, precancers do not appear until long after the animals are exposed to carcinogens. In the interim, we refer to a biological latency period, during which the cells with no discernible morphologic or genetic lesions develop into precancers. The latency period that precedes precancers is a biological mystery waiting to be solved.

In most cases, early precancers accumulate additional morphologic and genetic changes that increase over time. Countless studies have shown that the lowest grade precancers of a specific type have less atypia and fewer cytogenetic changes than more advanced precancers (58–61). By the time that a precancer is ready to become a cancer, it has already accumulated a complex set of changes. If we cannot accurately measure the moment when a precancer becomes a cancer, we no more invalidate the existence of the precancer than we invalidate the existence of the cancer.

Argument 3.

There are many genetic and morphologic disparities among different recognized precancers. Because these lesions seem to have no properties in common, other than the defining property of cancer precedence, why bother assigning them any biological status?

Response. As discussed in Chapter 4, there are many diverse types of precancers. Squamous dysplasia of the uterine cervix, tubular adenoma of colon, Barrett esophagus, myelodysplasia, and nephrogenic rests are all types of precancers, but they seem to be biologically unrelated.

Indeed precancers fall into diverse biological classes. So what? This does not mean that precancers do not exist or that you cannot study the biology of precancers. Because precancers come in different diverse biological forms, creating a biological categorization of precancers is necessary, so that the types of precancers with similar phenotypes can be studied together (65). The need to fund a classification effort for precancers was discussed in Chapter 5.

8.3 Irrelevance Arguments

Argument 4.

Since 1991, the cancer death rate has been steadily dropping, and cancer survival for many of the most common cancers has been increasing. Our successes in the war against cancer reassure us that we should not abandon our winning strategy by switching to precancer research.

Response. Many of the greatest advances in cancer survival result from improved detection of easily treatable, early cancers. As discussed in Chapter 1, our ability to cure the common, advanced cancers has not changed much over the years. At this time, there is no reason to believe that cures for the most common, advanced cancers are likely to be found any time soon.

Most people look at the bright side of life. Survival rates for many types of cancer are improving, particularly when cancers are detected early. Scientists, however, should not be cheerleaders for the public. Every cancer researcher is obliged to assess data dispassionately and to base one's actions on facts, not hope. The fact is that, over the past half-century, funded efforts to treat the common advanced cancers have not reduced the number of cancer deaths in adults. Abandoning the current approach for another would be a reasonable step, if there were sufficient evidence to believe that the alternate approach might succeed.

Argument 5.

The mission of the National Cancer Institute, the primary funding agency for cancer research in the United States, is to develop cures for cancer, not precancer. If precancers were as important as you say they are, there would be a National Precancer Institute; but, there isn't.

Response. This argument turns the tables on those who insist that precancer is a lesion that is distinct and separable from cancer. If precancers really are a different lesion from cancer, then why should precancers receive research funds earmarked for cancer? Well, the answer is obvious. The job of the National Cancer Institute is to eliminate cancer through research. Pursuing precancer research is the best strategy to eliminate cancer.

Argument 6.

Precancers regress spontaneously. Why should we try to develop treatments for a disease that usually resolves without treatment?

Response. Yes, many precancers regress spontaneously, and if we were to treat all the precancers, we would be treating many lesions that would have regressed without treatment. At this time, we cannot distinguish precancers that will regress from precancers that will progress to invasive cancer. Until we can distinguish regressing precancers from progressing precancers, we must treat them all.

At this point, we know almost nothing about the causes of precancer regression. We could potentially cause all precancers to regress, if we knew how to control the conditions that favor regression. If we could arrest the transition of precancer to cancer, then we could halt the occurrence of invasive cancers. If we could simply delay the transition of precancers to cancer, even if it were for just a few years, then we could greatly reduce the burden of cancer in the population.

8.4 Priority Arguments

Argument 7.

Precancer research falls under cancer prevention. This is true because when you treat a precancer, you prevent a cancer. Cancer prevention is an adequately funded area of cancer research, so we really do not need to assign any special funding to precancer research.

Response. Prevention is a funded research area at the U.S. National Cancer Institute, within the Division of Cancer Prevention. Cancer prevention involves identifying and eliminating carcinogens as well as adopting lifestyles that are thought to minimize exposure to carcinogens or to increase the ingestion of anticarcinogens (present in fruits and vegetables) in the diet. These priorities are not really aimed at precancer detection, diagnosis, and treatment.

Likewise, the precancers have not been included in the initiatives of NCI's Division of Cancer Treatment and Diagnosis, most of which are aimed at finding or testing new chemotherapeutic agents for cancers. Precancer treatment is an area that has not fallen neatly into any of the National Cancer Institute divisions. Precancers have never gotten the attention they actually deserve.

Argument 8.

People are dying from malignant tumors every day. Even if we could stop the occurrence of new cancers by treating the precancers, we cannot deviate from our commitment to relieve the suffering of those people whose cancers have progressed beyond the precancer stage.

Response. Individuals with advanced, metastatic cancers may or may not benefit from an expansion of research activity in the precancers. At this point, we simply do not know whether agents that treat precancers will have activity against metastatic or invasive tumors. Curing the commonly occurring advanced, metastatic cancers has been an intractable problem.

Many patients suffering from metastatic cancer wish they had lived in a world where cancers were stopped before they invade and metastasize. Precancer research will help create a better world for their loved ones.

Argument 9.

If precancers were as important as you seem to insist, then the major cancer institutions would have thrown in their support by now.

Response. The importance of the precancers was recently affirmed by the American Association for Cancer Research Task Force on the Treatment and Prevention of Intraepithelial Neoplasia (149). In this report, the Task Force recognized intraepithelial neoplasia (IEN) as a near-obligate precursor to invasive cancer and identified IEN as a treatable disease.

> *Reducing IEN burden, therefore, is an important and suitable goal for medical (noninvasive) intervention, to reduce invasive cancer risk and to reduce surgical morbidity. Achieving the prevention and regression of IEN confers and constitutes benefit to subjects and, in the opinion of this Task Force, demonstrates effectiveness of a new treatment agent.*

In addition, the NCI has recently launched a precancer initiative, described in detail in Chapter 6 (131).

8.5 Impracticality Arguments

Argument 10.

Treating precancers is not feasible for the majority of precancerous lesions that occur in humans. Reducing the incidence of cervical cancer by treating cervical precancer was possible only because the cervix can be inspected and sampled. There is no equivalent method to find and excise precancerous lesions of pancreas, lung, prostate, and breast. Procedures to detect and treat most precancers, therefore, are not practical.

Response. As we discussed in Chapter 7, it is wrong to think that precancers must be detected and diagnosed prior to treatment. Current efforts to prevent, eradicate, or prolong precancers, without requiring detection or diagnosis, can very likely be expanded in the future.

Argument 11.

There are thousands of cancers. Likewise there are thousands of precancers. Most precancers are unknown. It would be an impossible task to identify, collect, and study all the different human precancers.

Response. Although there are thousands of different cancers, the number of precancers is likely much smaller. As discussed in Chapter 5, a single precancer may progress to any of several different cancers. Furthermore, by classifying precancers into a

List 8.5.1 Economic advantages of treating precancers.

1. Proven to work in practice (low-risk concept)
2. Current research is highly successful (expected high return on investment)
3. Does not require any great breakthrough (low-risk research and development budget)
4. Rapid and relatively inexpensive carcinogenicity trials
5. Rapid and relatively inexpensive treatment trials
6. Can potentially eradicate cancer within the lifetime of most currently living Americans (high pay-out)

relatively small number of classes, the task of developing treatments for all precancers may be simplified. For example, all the intraepithelial neoplasms may fall under a single class of precancers, and all may be treatable by one general class of agents. Currently, we cannot say with any confidence that developing effective treatments for classes of precancers is possible. To test our hypothesis, we will first need to develop a precancer classification.

Argument 12.

We simply cannot afford to treat the precancers. The United States spends more money on health care than any other country on earth. Adding precancers, diseases of disputed existence, to the list of expensive American afflictions, is just impossible to contemplate.

Response. Precancer research competes with many other promising strategies for a share of the total cancer research budget. Still, the precancers offer some outstanding economic incentives (List 8.5.1).

8.6 Funding Priorities

Precancer research has paid off, in more lives saved, than any other area of cancer research. As we discussed in Chapter 1, every country that has deployed Pap smear screening has enjoyed a 70–90% drop in the number of cancer deaths from cervical cancer, the most common cause of cancer deaths in populations that do not use Pap smear screening. No other cancer intervention can match this accomplishment. A recent report issued by the National Cancer Institute concluded that "evidence suggests that 90% or more of colorectal cancer deaths could be prevented if precancerous polyps were detected with routine screening and removed at an early stage (1)." In 2008, 49,960 Americans died of colon or rectal cancer (2). A 90% reduction in the U.S. colon cancer death rate would save 45,000 lives each year.

List 8.6.1 Priorities for precancer research.

1. Begin testing nontoxic agents very soon on high-risk patients and patients known to have precancers (e.g., angiogenesis inhibitors, growth inhibitors, apoptotic agents, retinoids).
2. Begin testing toxic agents very soon that have already undergone Phase I, II, and III trials for anticancer activity. Potential precancer therapies can use agents previously developed and tested against cancers. Because such agents have already undergone Phase I, II, and III clinical trials, testing these same agents against precancers would be a sensible and economical strategy. Because precancers share many biological properties of cancers, hoping that agents that show activity against cancers will also show activity against precancers is reasonable.
3. Develop and fund precancer initiatives aimed at answering fundamental questions of precancer genesis, molecular signatures, key biologic pathways, and transition to cancer.
4. Standardize precancer terminology, and develop a biology-based (not morphology-based) classification. Specifically, classify precancers into groups of similar biology that are likely to respond in a similar manner to specific classes of antiprecancer agents. This will be the basis for a separate effort to formally classify precancers, so that all researchers will have a standard set of criteria for diagnosing lesions and a standard vocabulary for naming the lesions.
5. Fund biorepositories for precancer tissues.

Additional precancer research may lead to the eradication of most human cancers. Prioritized areas for precancer research are found in List 8.6.1. They are all low-risk and relatively low-cost initiatives. Funding agencies that have chosen to take a pass on the precancers should reconsider their positions.

Chapter 8 Summary

Cancer research is a zero-sum game. Each year, the National Cancer Institute, as well as other cancer funding agencies, are provided with a research budget. Money spent on lung cancer research is money diverted from breast cancer research. Money spent on breast cancer research is money diverted from prostate cancer research.

The NCI staff determines how much money can be spent on each area of cancer research. Currently, there is no NCI division or program devoted to precancers. There is no established group of scientists who are pushing for precancer funding. There are no congressmen or members of the Executive Office who are insisting on new precancer initiatives. Perhaps the time has come to examine the arguments against precancer research and to conclude that the time to support precancer research has finally arrived.

This chapter contains a list of high-priority areas for precancer research. All these areas are low risk or no risk. Every item on the list, with the exception of item 5 (standardizing a precancer terminology and developing a biology-based classification) are currently being funded by the NCI and other cancer research agencies. I believe that the level of funding is not commensurate with the importance of the research. More scientists, in more laboratories, must become involved in the effort.

Developing a standard terminology and classification requires gathering expert pathologists and biologists and asking them to produce a consensus document. Without a standard way of diagnosing lesions, different scientists from different laboratories will assign the same name to different lesions or different names to the same lesion, and their published findings will not reconcile. Every laboratory's published results will be unrepeatable. The precancer field, at present, has no standardized vocabulary, a situation that will impede progress until it is corrected.

CHAPTER

The End of Cancer

I don't try to describe the future. I try to prevent it.

—Ray Bradbury

9.1 Background

The past is behind us. The present will be over in an instant. What happens next? This chapter suggests a few scenarios, all extrapolated from current events and trends in the cancer field.

9.2 Future 1, Pestilence and Plagues

The incidence of cancer worldwide continues to rise as environmental pollution, contamination of dwindling food and water supplies, the cumulative burden of nuclear waste material, and reduction of the ozone layer continues unabated. The molecular-targeted treatments all fail, as tumors quickly develop resistance to these new drugs. For purely economic reasons, advanced treatments for most cancers continue to lie outside the reach of most individuals in the world. Cancer, once a disease that occurred primarily in older people, shifts into younger populations, presumably because of the effect of early exposures, beginning in utero, to thousands of new industrial carcinogens introduced in the last decades of the twentieth century.

9.3 Future 2, Micromanaging Cancer

Scientists develop strong mathematical techniques that can assign cancers into groups by their **molecular signatures**. This technology requires meticulous preparation and measurement of the genes expressed in an excised cancer and comparison of the **profile** of its expressed genes against the profile of other cancers of the same histologic type. Although the technology is expensive, costing thousands of dollars for each test, these studies accurately determine the best treatment possible for cancer patients. Most analyses indicate that patients with advanced cancer can expect a three-month increase in life expectancy thanks to this new technology. Patients with cancers that have not metastasized and who fall into the most favorable treatment group, can avoid toxic chemotherapy based on the results of these tests. Early successes in gene profiling of tumors have spurred the National Cancer Institute to expand its clinical trial program to develop new therapeutic approaches for subgroups of cancer patients identified by the technique. These new treatment options require another decade of development for the most common causes of cancer death in mankind (lung, colon, breast, prostate, and pancreas). Treatment options for the less common cancers will likely become available over the next 30 to 80 years.

9.4 Future 3, Public Health and Lifestyle Enhancements

Scientists agree that cancer is a largely preventable disease. Lifestyle modifications that include regular exercise, cessation of smoking, reduction of air pollution, removal of carcinogenic contaminants in food and water, strict control of caloric intake, adoption of diets rich in fruits and vegetables, and dietary supplementation with a balance of antioxidants, vitamins, and minerals reduce the incidence of cancer, overall, by 20%.

9.5 Future 4, Massive Screening and Surveillance

The term *precancer* has been abandoned by the scientific community in favor of the more easily understandable term *early cancer*. Serologic screening techniques are developed to identify patients who are likely to harbor early cancers. Powerful and expensive in vivo imaging techniques are developed to find malignant cancers when they are very small. Endoscopic surgical techniques are developed to extirpate small cancers located almost anywhere in the body. A massive campaign has begun, in the Unites States to screen, detect, diagnose, and excise small cancers. Compliance is a problem. The new screening protocols have poor sensitivity, high false-positive rates, and high costs. Death rates from some of the more common cancers fall modestly. The screening programs are hailed a major success by their developers and financial investors. Detractors are many.

9.6 Future 5, Precancer Treatment

People awaken to the importance of precancers. Decades of precancer neglect has produced a generation of researchers and healthcare workers who are woefully misinformed on the subject. Precancer initiatives train cancer researchers to diagnose precancers in a uniform manner, so that researchers in different laboratories, working on different aspects of the same precancers, produce comparable data. The community of trained precancer researchers develop a consensus on the most important clinical and biological questions in the field. New findings from researchers and clinical trialists lead quickly to methods that prevent or treat the precancers. The cancer incidence rate drops precipitously, coincident with a dramatic drop in the cancer death rate. Life expectancy increases by several years.

Chapter 9 Summary

One of the most fascinating things about science is that many different approaches to a common array of problems are pursued simultaneously. War is an apt metaphor for scientific advancement. During World War II, scientists worldwide engaged in many efforts to win the war. Research was funded on radar, cryptography, jet aircraft, rocketry, biological weapons, powerful explosives, gunnery, electronic surveillance, improved shielding for armored vehicles, miniaturization, ballistics, naval architecture, and so on. The military leadership in the United States recognized that an atomic bomb was feasible. They gathered all the scientific talent, laboratory support, and financial resources to develop a U.S. bomb before the Nazis developed one of their own. All of these urgent projects proceeded simultaneously and helped shape today's world. Similarly, world governments fund a variety of healthcare projects that are proceeding simultaneously: the human genome project, stem cell research, nanotechnology, **epigenomics**, gene-targeted chemotherapies, genetic engineering, public health interventions, vaccine development programs, large clinical trials, and so on. All of these efforts will likely shape the way we live and die in tomorrow's world.

One of the purposes of this book is to influence the future. The premise of this book is that if the public were educated in the biology of precancers, there would be more effort in the field, and effective treatments for the different classes of precancers would quickly follow. Modest short-term advances in precancer treatment will rapidly reduce the cancer death rate. In the long term, the precancers are the best hope we have for the eradication of the disease known as cancer. Should we not include precancer funding among the research programs supported by the world's major cancer funding agencies?

APPENDIX

1 Cancer Numbers

Our perception of reality is determined, to a large degree, by the way we measure and count objects. In the realm of cancer research, we count the number of people who have cancer, measure their length of survival, and acquire additional information that stratifies the counts based on demographic or clinical measurements. We use available documents, such as official census counts, Medicare data (in the United States), and medical records. The quality of clinical research often depends on the quality of medical records.

When medical records are incomplete, incorrect, illegible, or otherwise uninterpretable, the results for an otherwise well-planned clinical trial can be disastrous. One source of crucial disease data is the death certificate, which provides a final statement of every individual's cause of death (150). Death certificate data have many deficiencies (151, 152). The most common error occurs when a mode of death is listed as the cause of death. For example, cardiac arrest is not a cause of death, although it appears as the cause of death on many death certificates. There is not much value in a death certificate for a person who died with end-stage cancer when the listed cause of death is "cardiac arrest." An international survey has shown very little consistency in the way that death data are collected (153). In the vast majority of cases, death certificates are completed without the benefit of an autopsy. At best, death certificates express a clinician's reasonable judgment at the time of a patient's death.

For this book, we have tried to limit ourselves to respected data sources. The U.S. federal government regularly undertakes large data collection efforts, employing armies of personnel over many years, during which their activities are continuously examined and improved. We made extensive use of data produced by the U.S. Census Bureau, the U.S. Center for Disease Control and Prevention, and the National Cancer Institutes Surveillance, Epidemiology and End Results Program. Nonetheless, numbers can be deceiving. Listed here are most of the "facts" included throughout the book, along with their sources. Readers are encouraged to be skeptical of these numbers and to verify their accuracy.

London deaths from cancer in the years 1651–1758 (excluding plague years) expressed as a percentage of deaths from all causes recorded in the *Yearly Bills of Mortality* (7): 0.26%

U.S. cancer death rate in 1871 (not age-adjusted) (7): 36.9/100,000

U.S. cancer death rate in 1881 (not age-adjusted) (7): 52.3/100,000

U.S. cancer death rate in 1891 (not age-adjusted) (7): 60.9/100,000

U.S. cancer death rate in 1901 (not age-adjusted) (7): 73.1/100,000

U.S. cancer death rate in 1911 (not age-adjusted) (7): 92.6/100,000

U.S. population in 1915 (154): 100,546,000

World population in 1915 (7): 1,750,000,000

U.S. deaths from cancer in 1915 (7): 80,000

U.S. cancer death rate (age-adjusted) in 1950 (8): 195.4/100,000

U.S. cancer death rate (age-adjusted) in 1975 (155): 199.1/100,000

World population in 2000 (156): 6 billion

Worldwide deaths from cancer in 2000 (157): 7,000,000

Worldwide deaths from cancer in 2000 expressed as a percent of deaths from all causes (157): 13%

U.S. lifetime risk of developing cancer (2002–2004) (5): 41%

U.S. cancer death rate (age-adjusted) in 2001–2005 (158): 189.8/100,000

U.S. cancer death rate (age-adjusted) in 2005 (155): 184.0/100,000

U.S. population in 2008 (156): 303,824,640

U.S. life expectancy (for a person born in 2008) (156): 78.14 years

U.S. median age (156): 36.7 years

U.S. deaths from cancer in 2008 (2): 565,650

U.S. death rate in 2008 (156): 827/100,000

U.S. new cancer cases in 2008 (excludes squamous cell carcinoma of skin, basal cell carcinoma of skin, and all precancers with the exception of bladder carcinoma *in situ*): 1,437,180

U.S. deaths from lung cancer in 2008 (2): 161,840

Percentage of U.S. cancer deaths from lung cancer in 2008: 28.6% (161,840/565,650)

U.S. deaths from colorectal cancer in 2008 (2): 49,960

Percentage of U.S. cancer deaths from colorectal cancer in 2008: 8.8% (49,960/565,650)

U.S. cancer death rate in 2008 (not age-adjusted): 186.2/100,000 (565,650 × 100,000)/303,824,640)

World population in July 2008 (estimate) (156): 6,706,993,152

APPENDIX

2 Precancer Terminology

The terminology for human precancers has not been standardized (65, 72, 116, 117). Some of the lesions called "precancers" in this book would be considered benign hyperplastic lesions, benign neoplasms, or even cancers by other pathologists. For example, an autonomously functioning hyperplastic nodule in the thyroid might be considered a thyroid precancer (not a hyperplasia) by some pathologists, or a benign neoplasm by others. A MALToma might be considered a malignant lymphoma by the majority of pathologists, but some pathologists, such as ourselves, consider it a prelymphoma. Even where there is general agreement that a lesion is a precancer, there may be no agreement on the proper name for the lesion or its relationship to different precancer stages leading to the same malignant neoplasm. Still, it is difficult to understand the relevant precancer literature without some exposure to the words that pathologists commonly apply to lesions that are often assumed to be precancers (List A.1).

Cancers arise from virtually every type of cell in humans and animals. For all common cancers of humans, and for many less common cancers, there is an identified precancer. We can list every anatomic region and some of the generally recognized precancers that occur at these sites (List A.2).

List A.1 For each tissue, commonly used precancer names are listed, ranging from the mildest recognized form of precancer up to the progression from precancer to invasive carcinoma.

For squamous epithelium

Squamous metaplasia*

Mild dysplasia Moderate dysplasia Severe dysplasia

Carcinoma *in situ* (stage 0 squamous cell carcinoma)

Microinvasive carcinoma

Invasive carcinoma

For glandular epithelium

Glandular metaplasia**

Glandular hyperplasia Florid glandular hyperplasia Glandular hyperplasia with atypia

Mild glandular dysplasia Moderate glandular dysplasia Severe glandular dysplasia

Adenocarcinoma *in situ* (stage 0 adenocarcinoma)

Microinfiltrating adenocarcinoma

Invasive adenocarcinoma

For ductal epithelium

Ductal hyperplasia Florid ductal hyperplasia Florid ductal hyperplasia with atypia

Atypical ductal hyperplasia Ductal intraepithelial neoplasia

Ductal carcinoma *in situ* (stage 0 adenocarcinoma)

Microinfiltrating ductal carcinoma

Infiltrating ductal carcinoma

For melanoma

Nevus Nevus with cytologic atypia Nevus with cytologic atypia and architectural disorder Dysplastic nevus

Melanoma *in situ*

Thin melanoma Thick melanoma

For prostate

Low-grade prostatic intraepithelial neoplasia

High-grade prostatic intraepithelial neoplasia

Prostatic adenocarcinoma

For cervix

Low-grade squamous intraepithelial lesion

High-grade squamous intraepithelial lesion

Cervical carcinoma *in situ*

Microinvasive carcinoma of cervix

Invasive carcinoma of cervix

*Applies to tissues normally devoid of squamous cells but in which squamous cells appear as a pathologic process. Examples: Focal replacement of the ciliated epithelium of the bronchus with squamous epithelium; focal replacement of the urothelial cell mucosa of the bladder with squamous epithelium.

**Applies to focal replacement of normal glandular or squamous epithelium with glandular epithelium that morphologically resembles the glandular mucosa from some other anatomic location. Examples: Focal replacement of the squamous mucosa of the esophagus with mucosa of the intestine; focal replacement of the glandular mucosa of the stomach with mucosa of the intestine.

List A.2 List of precancers (organized by anatomic site or system).

Oronasal cavities (lips, tongue, oral mucosa, nasal cavity and accessory sinuses, nasopharynx, pharynx, and hypopharynx, which includes salivary gland tumors, odontogenic tumors, and soft tissue tumors)

- Leukoplakia of lips
- Buccal mucosa leukoplakia
- Gingival leukoplakia
- Glossal leukoplakia
- Leukokeratosis of oral mucosa
- Nasopharynx carcinoma *in situ*
- Oropharynx carcinoma *in situ*
- Lip carcinoma *in situ*
- Epiglottis carcinoma *in situ*

Larynx (includes soft tissue tumors)

- Leukoplakia arising in vocal cords
- Larynx carcinoma *in situ*

Lung (includes trachea and bronchus and excludes primary tumors of visceral pleura)

- Bronchial metaplasia with squamous cell dysplasia
- Lung nonsmall cell carcinoma *in situ*
- Adenomatous hyperplasia
- Atypical adenomatous hyperplasia (AAH)
- Bronchioloalveolar carcinoma, pure lepidic form
- Atypical glandular hyperplasia of lung
- Lung with adenosquamous carcinoma *in situ*
- Lung with epidermoid cell carcinoma *in situ*
- Diffuse idiopathic pulmonary neuroendocrine cell hyperplasia (DIPNECH)
- Bronchus with squamous dysplasia
- Carcinoma *in situ* of trachea
- Carcinoma *in situ* of bronchus
- Ground glass opacity of lung (radiologic)

Heart and vascular system

Digestive system (includes soft tissue tumors arising in submucosa but excludes mucosal and submucosal lymphomas)

- Gallbladder with adenocarcinoma *in situ*
- Esophagus
  - Barrett esophagus
  - Dysplasia in Barrett esophagus
  - Esophageal squamous cell carcinoma *in situ*
  - Esophagus with adenocarcinoma *in situ*

Stomach
- Intestinal metaplasia of gastric mucosa
- Atypical intestinal metaplasia
- **Gastric dysplasia**
- Adenomatous polyp of stomach
- Gastric adenocarcinoma *in situ*

Small intestine (includes duodenum, ampulla of Vater, jejunum, and ileum)
- Adenomatous polyp of ampulla
- Small intestinal adenoma

Large intestine

Cecum
- Cecal adenoma
- Cecal adenocarcinoma *in situ*

Appendix
- Appendiceal adenoma
- Appendiceal carcinoma *in situ*

Colon and Rectum
- Adenocarcinoma *in situ* in tubular or tubulovillous adenoma
- Adenocarcinoma *in situ* in villous adenoma
- Colon with adenocarcinoma *in situ*
- Colon with adenoma
- Colorectal adenocarcinoma *in situ*
- Aberrant crypt of colon

Anus
- Anal leukoplakia
- Anal intraepithelial neoplasia
- Anal margin Bowen disease

Liver, gallbladder, and bile duct
- Liver dysplasia (61)
- Dysplasia of extrahepatic bile duct
- Carcinoma *in situ* of extrahepatic bile duct

Pancreas
- Exocrine pancreas
  - Atypical papillary hyperplasia of pancreatic ducts
  - Pancreatic intraepithelial neoplasia (PanIN) (67)
  - Ductal carcinoma *in situ*
  - Intraductal papillary carcinoma of pancreas
  - Intraductal papillary-colloid adenocarcinoma
- Endocrine pancreas (pancreatic islets)

(continued)

List A.2 Continued

Pleural/peritoneal surfaces (includes pelvic surfaces and all visceral peritoneal surfaces, except for ovarian surfaces)

Spleen (excludes lymphomas and leukemias of spleen)

Eyes and orbit (excludes optic nerve)

Central nervous system (brain coverings, leptomeninges and dura, and soft tissue arising intracranially, but excludes pituitary)

Skin (excludes lips, but includes ears and other skin of face; includes melanomas occurring on skin, epithelial tumors of epidermis and appendages; excludes primary lymphoid lesions of skin; includes soft tissue tumors of dermis and subcutis; includes nonareolar skin overlying breast, but otherwise excludes all lesions of breasts and areola)

- Arsenical keratosis
- Actinic keratosis
- Skin carcinoma *in situ*
- Skin squamous cell carcinoma *in situ* (Bowen disease)
- Melanocytes and nevus cells
    - Dysplastic nevus
    - Lentigo maligna, precancerous melanoma
    - Melanoma *in situ* (superficial spreading, nodular, lentigo maligna, and acrolentiginous subtypes)

Bones and joints (includes vertebrae and cartilage, but excludes bone marrow)

- Enchondroma
- Ostechondroma
- Fibrous dysplasia

Bone marrow (includes hematopoietic tissue within bone, but excludes extramedullary hematopoietic tissue, such as spleen)

- Myeloproliferative disorders
    - Polycythemia vera
    - 8p11 myeloproliferative syndrome
    - Essential thrombocythemia
    - Idiopathic hypereosinophilic syndrome
    - Systemic mastocytosis
    - Monoclonal gammopathy of undetermined significance
- Melodysplastic syndromes
    - Refractory anemia (erythrodysplasia)
    - Refractory anemia with ringed sideroblasts
    - Refractory anemia with excess blasts
    - Refractory anemia with excess **blast transformations**
    - Myelodysplastic syndrome with monosomy 7

Mixed myeloproliferative-myelodysplastic syndrome
Myelofibrosis

Soft tissue (excludes soft tissue of named organs/anatomic sites, but includes soft tissue in extremities and trunk exclusive of skin)

Breast

Fibrocystic disease of breast with atypical hyperplasia
Florid epithelial hyperplasia of breast
Flat epithelial atypia (160)
Microglandular adenosis (62)
Aypical apocrine metaplasia
Atypical ductal hyperplasia
Atypical lobular (glandular) hyperplasia
DIN (ductal intraepithelial neoplasia)
Breast LCIS (lobular carcinoma *in situ* of breast glands)
Breast DCIS (ductal carcinoma *in situ* of breast ducts)
Breast comedo DCIS (necrotic)
Breast noncomedo DCIS

Gynecologic sites exclusive of breast (excludes germ cell tumors, but includes stromal tumors of these sites)

Ovaries and appendages (includes ovarian surfaces and excludes ovarian germ cells)
Ovarian intraepithelial neoplasia (OIN)
Borderline ovarian tumor
Uterus
Corpus
Endometrial intraepithelial neoplasia (EIN)
Endometrial adenocarcinoma *in situ*
Myometrium
Seedling myoma
Cervix
Squamous intraepithelial lesion (cytologic diagnosis)
Atypical glandular cells (cytologic diagnosis)
Adenocarcinoma *in situ* of cervix (cytologic and histologic diagnosis)
Cervical intraepithelial neoplasia 1, 2, or 3 (histologic diagnosis)
Vagina
Vaginal intraepithelial neoplasia
Leukoplakia arising in vagina
Vulva (includes accessory structures and perineum)
Vulvar intraepithelial neoplasia
Vulvar lichen sclerosis et atrophicus (kraurosis vulvae)

(continued)

List A.2 Continued

Male genitalia and accessory structures (includes testes, but excludes germ cells)

- Penis (includes penile urethra)
  - Bowenoid papulosis of penis
  - Lichen sclerosis et atrophicus (kraurosis penis)
  - Erythroplasia of Queyrat (Bowen disease of penis)
- Scrotum
- Testes (excludes germ cell tumors)
- Prostate, seminal vesicles, vas deferens, and prostatic urethra
  - Prostatic intraepithelial neoplasia
  - Low-grade prostatic intraepithelial neoplasia
  - High-grade prostatic intraepithelial neoplasia
  - Intraductal dysplasia (63)
  - Prostatic adenocarcinoma *in situ*

Kidney and ureters (includes urothelial cell lesions, excludes embryonic tumors)

- Renal adenoma
- Urothelial cell carcinoma *in situ* of renal pelvis and ureters
- Urothelial cell dysplasia of renal pelvis and ureters
- Papillary urothelial cell tumor grade 0 of renal pelvis and ureters

Urinary bladder

- Urothelial carcinoma *in situ* of urinary bladder
- Flat carcinoma *in situ* of urinary bladder
- Urothelial cell dysplasia of urinary bladder
- Papillary urothelial cell tumor grade 0 of urinary bladder
- Squamous metaplasia of urinary bladder

Lymphoid tissue (includes lymph nodes, spleen, and lymphoid collections in any organ, including gastrointestinal tract, respiratory tract, and skin)

- Angiocentric immunoproliferative lesion
- Angioimmunoblastic lymphadenopathy
- Angioimmunoblastic lymphadenopathy type T-cell lymphoma
- Angioimmunoblastic lymphadenopathy with dysproteinemia
- Lymphomatoid granulomatosis
- Pulmonary lymphomatoid granulomatosis
- Cutaneous lymphomatoid granulomatosis
- Lymphoproliferative disease (excludes immunoproliferative disorders of myeloid origin)
- Waldenstrom macroglobulinemia
- Autoimmune lymphoproliferative disorder

Lymphoproliferative disorder X-linked
Atypical lymphoid hyperplasia
Follicular lymphoma *in situ*
MALT lymphoma (MALToma)
Posttransplant lymphoproliferative disorder
Cutaneous prelymphoma
Cutaneous T-cell lymphoid dyscrasia (161)
Indeterminate lymphocytic lobular panniculitis
Atypical lymphocytic lobular panniculitis
Premycosis fungoides
Parapsoriasis en plaque
Poikiloderma atrophicans vasculare or prereticulotic poikiloderma
Woringer-Kolopp disease, localized epidermotropic reticulosis
Actinic reticuloid
Pagetoid reticulosis
Generalized pagetoid reticulosis
Pagetoid reticulosis, Ketron-Goodman variant
Follicular mucinosis, folliculocentric mycosis fungoides

Endocrine organs (includes pituitary, thyroid, parathyroid, parafollicular [C cell] thyroid, adrenal [medullary and cortical], and gonadal endocrine)

Focal hyperplasia

Nodular hyperplasia

Adenoma

Atypical follicular adenoma of thyroid (159)

Germ cells

Intratubular germ cell neoplasia

Embryonic rests

Nephrogenic rests
Perilobular nephrogenic rests
Intralobular nephrogenic rests
In situ neuroblastoma (113)

Placenta and extraembryonic gestational tissues

Hydatidiform mole
Complete mole
Partial mole
Placental site trophoblastic tumor

There are many acquired and inherited diseases that carry an increased risk for developing precancers (Lists A.3 and A.4).

As a general rule, we have the greatest number of designated precancers for those tumors that are the easiest to inspect visually. Because the easiest tumors to inspect are the tumors that arise from skin, we have many different names for skin cancers, even if those cancers are rare. Squamous cell carcinoma of skin, melanoma, and **mycosis fungoides** (a lymphoma arising in the skin) all have abundant identified precancers. Tumors that arise from internal organs (e.g., brain, endocrine glands, connective tissue, gonads, lymph nodes) have very few designated precancers.

List A.3 Acquired precancerous conditions.

Acquired immunodeficiency syndrome (AIDS)
Paget disease of bone
Ulcerative colitis
Acquired lymphedema
Stewart-Treves post-mastectomy lymphedema
Aplastic anemia
Atrophic gastritis
Cirrhosis of liver

List A.4 Inherited precancerous conditions.

Diffuse palmoplantar keratoderma (tylosis)
Milroy disease, hereditary lymphedema
Fanconi syndrome
B-K mole syndrome
Familial adenomatous polyposis
Xeroderma pigmentosum
Cyclic neutropenia

APPENDIX

Neoplasm Classification

3

A comprehensive terminology for precancers is available from within the Developmental Lineage Classification and Taxonomy of Neoplasms (Figure A.1). This neoplasm classification contains nearly 130,000 names of neoplasms, designating about 6,000 different neoplasm concepts under 40 different classes and subclasses. Each precancer concept is labeled as such and listed with its synonyms. The schema for the classification is shown here and fully explained in the author's publication, *Neoplasms: Principles of Development and Diversity* (110).

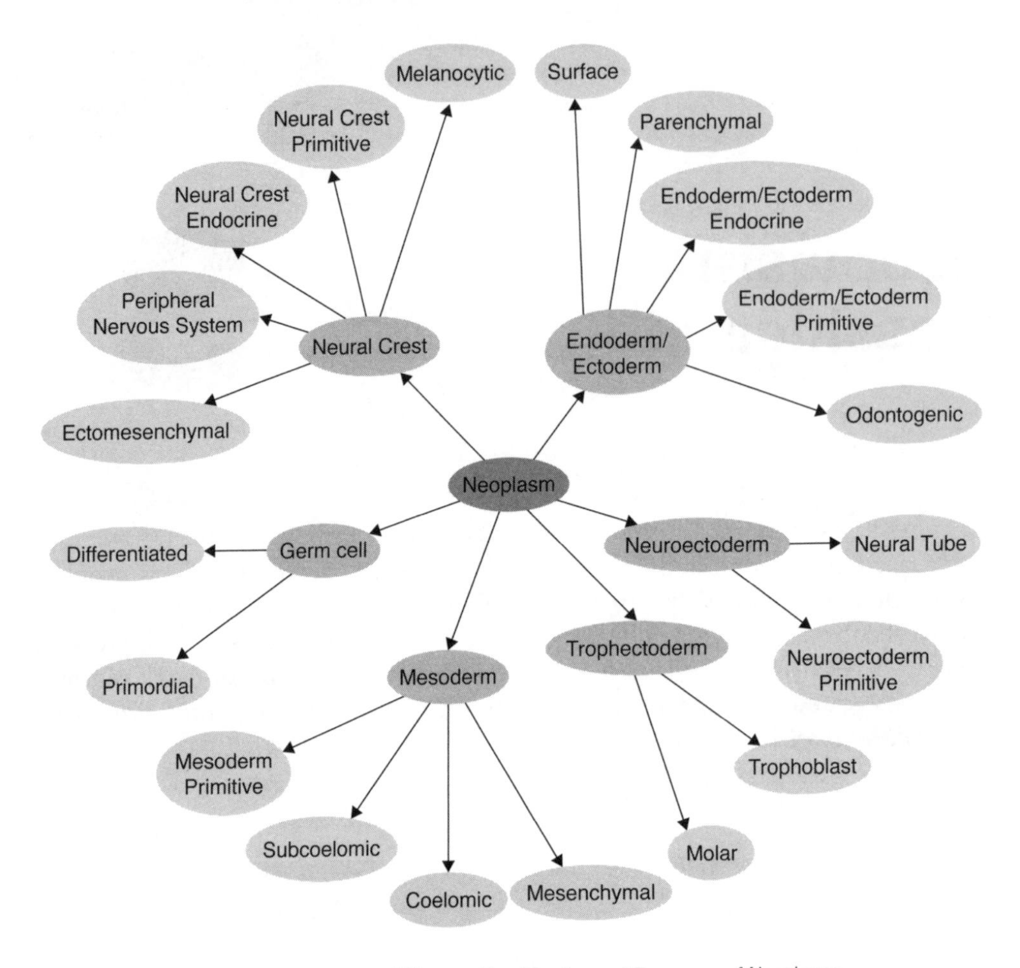

FIGURE A.1 Schema for the Developmental Lineage Classification and Taxonomy of Neoplasms.

The Developmental Lineage Classification is available at no cost as a curated, open-source document from the author's Web site http://www.julesberman.info/devclass.htm. The complete list of 6,001 precancer terms, organized under 229 precancer concepts can be obtained by visiting the search engine http://www.julesberman.info/neoget.htm and entering the word "precancer" (Figure A.2).

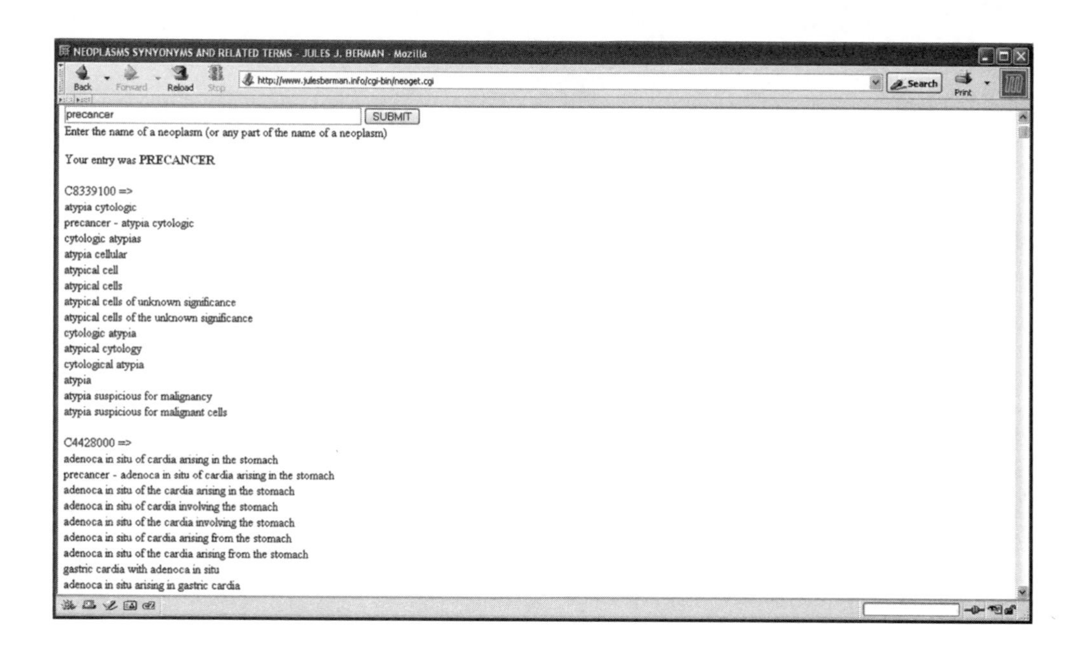

FIGURE A. 2 Neoplasm classification query page, with partial output for "precancer" entry.

Glossary

Ab initio—Latin: from the beginning.

Accrue (noun, accrual)—In the context of clinical trials, accrual is the recruitment of patients into the study. Accrual is one of the most difficult aspects of clinical trial management.

Actinic keratosis—One of the most common precancers, actinic keratosis forms on sun-exposed skin, hence, its alternate name, solar keratosis. Actinic keratoses occur as rough, crusted, and sometimes erythematous small areas of skin. Histologically, they are characterized by atypical squamous cells confined to the epidermis. On rare occasions, usually after the actinic keratosis has been left untreated for years, it may progress to an invasive lesion, at which time it becomes a squamous cell carcinoma. Squamous cell carcinomas that develop from actinic keratoses are unlikely to metastasize to distant organs.

Acute lymphocytic leukemia—Also known as acute lymphoblastic leukemia (ALL). ALL can occur at any age, but it occurs most frequently in children. Most children with ALL are cured of the disease. About 30% to 40% of adults with ALL achieve a long-term survival, compared with 80% in children.

Acute promyelocytic leukemia—Neoplastic disease of early-form white blood cells (promyelocytes). Acute promyeloctyic leukemia is one of the few neoplasms that can be successfully treated with an agent (all-trans-retinoic acid) that operates on the epigenome to induce differentiation. In combination with other drugs, about 90% of patients with acute promyelocytic leukemia achieve a clinical remission.

Adenocarcinoma—A cancer derived from glandular epithelium.

Advanced cancer—As used in this book, an advanced cancer is a cancer that has invaded extensively at its site of origin (i.e., into adjacent organs or into large vessels) or that has metastasized to other sites. Surgeons have been fairly successful at curing cancers that are not advanced (i.e., cancers that can be completely resected at their primary sites of growth, before they have metastasized). We have not had great success in curing the common cancers of adults (lung, breast, prostate, colon, pancreas, esophagus, liver, and so on) when they occur in their advanced stages.

Age-adjusted incidence—Over time, the age distribution of a population may change. Most human diseases are more likely to occur at specific age groups (e.g., the incidence of measles and chickenpox is higher in children than adults). A shift in the age distribution of a population can produce profound changes in the incidence of diseases, with no underlying biological cause. Because cancer is a disease that disproportionately effects seniors, an upward shift in the proportion of older people in a population will result in an observed rise in the incidence of cancer. To compensate for distortions in population data due to age-shifting, statisticians "adjust" incidence data by normalizing incidence and mortality data to a standard population distribution. Using age-adjusted rates when assessing clinical trends in the incidence or mortality of cancer is important. If, however, you want to assess the burden of cancer on a population, you need to know the crude (unadjusted) numbers for cancer incidence, cancer mortality, and the total number of new cancer cases.

Anatomic pathologist—Anatomic pathologists are physicians who have been trained to examine and evaluate lesions by their gross and microscopic appearances. Anatomic pathologists perform autopsies and render diagnoses on tissue biopsies. They provide intraoperative consultations to surgeons by performing rapid analyses of quick-frozen sections of sampled tissues removed during surgery. All cytopathologists have received training in anatomic pathology. All forensic pathologists (physicians who perform autopsies for the Medical Examiner's Office) have received training as anatomic pathologists. See *Glass slides.*

Anemia—Reduction in the number of circulating normal red blood cells.

Aneuploid—Abnormal number of chromosomes in a cell. See *Euploid.*

Angiogenesis—Process by which new vessels grow; synonymous with neovascularization. Angiogenesis is an important concept in cancer, because invasive tumors must somehow induce new vessel formation to grow progressively. The most common mechanism of tumor angiogenesis involves the secretion of endothelial growth factors by tumor cells, inducing new vascular growth response in the nonneoplastic stroma (connective tissue) that intermingles with the tumor. Precancers are *in situ* lesions that replace the normal epithelium, without invading into connective tissue. Consequently, most precancers do not require angiogenesis for their survival. One of the contemplated methods for treating precancers involves treating patients with angiogenesis-inhibiting agents, which may act by inhibiting the progression of precancers into invasive cancers. The idea is that the acquisition of invasion involves the acquisition of angiogenesis. If angiogenesis is inhibited, then invasion will not occur.

Aplastic anemia—Anemia caused by collapse of the population of cells in the bone marrow that produces circulating blood cells. Aplastic anemia is a life-threatening condition. Survivors of aplastic anemia are at increased risk for developing leukemia.

Apples-oranges bias—Clinical trials are carefully designed projects that follow protocols for every aspect of the trial, including source and preparation of drugs, doses and routes of administration, times and frequencies of administration, test groups, data collection techniques, and on and on. Even when two clinical trials seem comparable, they may be asking different sets of questions or using different analysis tools. Comparing survival data from two different clinical trials is a difficult analytic process, with many potential pitfalls.

Atypia—Deviation from normal cellular morphology. In the context of this book, atypia refers to changes in the morphology of a cell nucleus, indicating that the cells have become precancerous or cancerous. Atypia that is diagnostic of precancer is called "dysplasia." See *Dysplasia.*

Atypical—Having atypia.

Atypical adenomatous hyperplasia (AAH)—Any hyperplastic process of glands involving morphologic atypia. Conventionally reserved for a precancer of lung that may precede the development of a particular adenocarcinoma that arises from the bronchioles (small branches of the bronchi) and alveoli (air-exchanging sacs) (162).

Atypical intraductal hyperplasia (AIDH)—Same as atypical ductal hyperplasia (ADH), a form of ductal intraepithelial neoplasia (DIN), with mild atypia. DIN is the precancer for infiltrating ductal carcinoma of breast.

Autopsy—Also known as postmortem examination and necropsy.

Basement membrane—Thin membrane that anchors an epithelial surface to its underlying connective tissue. The basement membrane is a matrix of glycoproteins, collagen, and assorted macromolecules. On hematoxylin and eosin stain, the basement membrane is seen as a thin, pink line under the epithelium.

Benign—Not harmful. In the context of neoplasms, a benign tumor is an expanding clonal lesion that neither invades nor metastasizes. Some benign tumors, such as meningiomas in the brain, are nonetheless "harmful," because they push against surrounding vital structures. Benign tumors may occasionally transform into malignant tumors. The frequency of malignant transformation varies with the type of benign tumor. In general, the larger the benign tumor, the greater the likelihood that it may produce a focus of malignant cells.

Benzene—Ring-shaped aromatic hydrocarbon derived from petroleum and used in many different products. The effects of benzene are greatest in the bone marrow, where it causes aplastic leukemia, myelodysplasia, and leukemias.

Blast cells—A term that usually refers to dividing cells of the myeloid (blood cell) lineage.

Blast transformation—Synonymous with blast crisis. A rapid transformation from chronic myelocytic leukemia (in which the vast majority of circulating neutrophils are mature cells) to a phase in which the blood and the bone marrow are dominated by blast cells, creating a clinical and pathologic picture resembling acute myelocytic leukemia.

Bone marrow—Bone marrow contains adipose tissue and myelopoietic (blood cell producing) tissues. The blood cells produced in the marrow include red blood cells, thrombocytes (platelets), granulocytes (neutrophils, eosinophils, basophils), lymphocytes, monocytes, lymphocytes, and plasma cells. Most leukemias (neoplasms of circulating blood cells) arise from the bone marrow, and any type of myeloid cell or myeloid precursor may dominate in the resulting varieties of leukemia.

Bronchogenic carcinoma—Any cancer that arises from the bronchi. The most familiar bronchogenic carcinomas are squamous cell carcinoma, adenocarcinoma, and small cell carcinoma. Large cell carcinoma, a type of poorly differentiated carcinoma composed of large cells that have occasional features of squamous, glandular, and neuroendocrine cells, is usually included as a bronchogenic carcinoma. In addition, several rare cancers are bronchogenic. The term *bronchogenic carcinoma* excludes cancers that arise from alveoli and terminal bronchioles, particularly bronchioloalveolar carcinoma.

Burkitt lymphoma—Type of lymphoma that has a characteristic translocation that enhances activity of the c-MYC oncogene.

Carcinogen—Chemical agent, virus, or physical agent, such as ultraviolet light, that can initiate carcinogenesis.

Carcinogenesis—Biological process that leads to cancer. Carcinogenesis has the following consecutive stages: initiation, latency, precancer, and emergence of invasive cancer. Once cancer arises, the carcinogenesis stage ends, and the progression stage begins.

Carcinoma—Greek: *karkinos* meaning crab. Cancer that arises from epithelial tissue.

Carcinoma in situ—Noninvasive carcinoma. Because this carcinoma is not invasive, it is localized to its original place or origin (hence, *in situ*, or in place). Carcinoma *in situ* is the last phase in the development of a precancer. Once invasion begins, the lesion is no longer a carcinoma *in situ* and no longer a precancer. If we knew how to stop the progression from carcinoma *in situ* to invasive carcinoma, we could eliminate most deaths from cancer.

Cell division—Cell replication through the process of mitosis.

Cell morphology—Features of a cell seen through a microscope.

Cell size—Cells vary enormously in size and shape. A neuron can stretch, by axonal extension, to several feet in length! Human oocytes are many times larger than the size of most other cells (through a phenomenon called anisogamy). Regardless of the size of the cell, the nucleus of a diploid human cell (cell with a full chromosome complement) varies little among celltypes. Most normal nuclei have a diameter of about 7 to 10 microns (millionths of a meter). One of the most reliable distinguishing features between precancer cells and cancer cells is enlargement of the nucleus. A precancer nucleus can have several times the diameter of a normal nucleus, and it would be very unusual for a precancer nucleus to be as small as a normal nucleus.

Cell type—There are different kinds of cells in any animal organism. Examples of specific cell types are lymphocyte, squamous cell, neuron, and oocyte. There are at least 200 major cell types in the human body. The developmental process whereby cells achieve a specific mature cell type is called differentiation.

Cherry-picking—Choosing the data that suit your agenda, while omitting data that would jeopardize your conclusions or cause some obstacle or inconvenience in your research plans, or otherwise not suit your agenda. Cherry-picking is a method for producing bias in a clinical trial or experiment.

Childhood cancer—Cancers of childhood are rare. Leukemias and brain tumors account for the majority of new cases. Other tumors include neuroblastoma, Wilms tumor, rhabdomyosarcoma, and osteosarcoma. The cancers that occur most commonly in children are rare or uncommon (in the case of acute lymphoblastic leukemia) in adults. The common cancers of adults (e.g., adenocarcinoma of lung, colon, pancreas, breast, and prostate) are extremely rare in children. When adult-type tumors occur in children, the tumors may have a distinctive clinical and genetic marker distinguishing it from tumors of the same tissues occurring in adults (e.g., juvenile secretory carcinoma of breast and midline carcinoma of lung occurring in young persons). Many of the greatest successes in cancer chemotherapy have involved rare childhood tumors, whereas the common advanced cancers of adults have defied decades of determined efforts to develop cures. Although the cure rate of childhood cancer has improved enormously in the past few decades, so has the incidence. SEER data indicate that among children under 14 years of age, between 1974 and 1991 there has been a 1% average yearly increase in the incidence of all malignant neoplasms combined (163). In general, the overall incidence of childhood cancers is rising, while the overall death rate has been falling (Figure G.1).

Choriocarcinoma—Rare tumor derived from trophoblastic cells (normally present in the placenta).

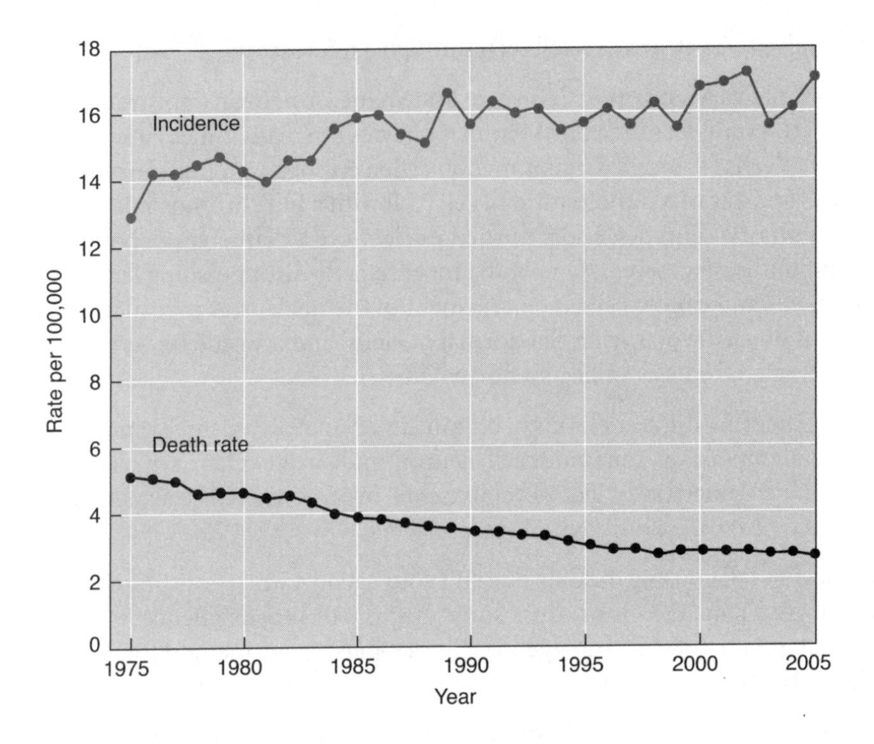

FIGURE G.1 The age-adjusted incidence rate and death rate for all cancers in children under the age of 20, expressed per 100,000 persons. Data from SEER Fast Stats. Statistics Stratified by Data Type. http://seer.cancer.gove/factstats/selections.php.

Chromatin—Chromosomal material (DNA + chromosomal proteins) that is stained blue with routine hematoxylin and eosin stain. Because standard histologic stains preferentially bind to protein, not DNA, the visualized nuclear material represents a textured amalgam of proteins attached to DNA and proteins in the nuclear matrix. In normal cells, chromatin is evenly dispersed through the nucleus. In cancer cells, chromatin appears as irregularly sized and shaped blue areas, separated by unnaturally light areas (parachromatin clearing). Abnormalities in chromatic distribution, a hallmark of dysplastic (precancerous) and cancerous cells, result from profound changes in the genetic and epigenetic organization of chromosomes.

Chronic myelogenous leukemia (CML)—Also known as chronic myelocytic leukemia and chronic granulocytic leukemia. A blood cancer characterized by abnormally high numbers of circulating granulocytes. In almost all cases of CML, there is a characteristic fusion gene (BCR/ABL) resulting from reciprocal translocation between chromosomes 9 and 22, the so-called Philadelphia chromosome marker of CML.

Cilia—Many terrestrial organisms contain fine, hair-like cell protrusions that wave back and forth. This wave movement can propel a cell, propel food toward the cell, or propel objects away from the cell. In prokaryotes (bacteria), the protrusion is called a flagellum. In eukaryotes (including plant sperm cells, protoctists, and some animal cells), the protrusion is called a undulipodium. Flagella are structurally different from undulipodia, but all undulipodia in eukaryotes are complex structures that share some common features. In humans, the cilia of epithelial cells and the tail of sperm cells are types of undulipodia. The cilia of respiratory lining cells work in unison, to flush mucus and particulate matter up and out of the tracheobronchial tree.

Classification—Hierarchical, comprehensive grouping of all members of a well-defined knowledge domain.

Clinical trial—Experiment using human subjects to test a clinical hypothesis. In many instances, clinical trials determine the effect of a drug or treatment protocol on patients with a specific disease or stage of disease. Today, all potential cancer cures are tested in clinical trials. Most cancer clinical trials are randomized clinical trials. Randomization is the random assignment of patients to treated versus untreated (or standard treatment) groups. In addition to being randomized, many clinical trials are double-blinded, so that neither the treating physician nor the patient know to which assignment group they belong during the trial.

Comorbidity—Two or more simultaneously occurring diseases in a patient.

Colon—As used in this book, the colon is equivalent to the large intestine and consists of the entire length of the large intestine, beginning with the cecum (which begins at the ileocecal valve) and extending to the anal canal. As used here, the colon includes the rectum. Anatomists and surgeons would argue that the colon and rectum are anatomically distinct and prefer to use the term *colorectal* when referring to conditions that can occur in either organ.

Colonoscopy—Procedure by which a flexible fiberoptic tube is threaded through the entire length of the colon. The colonoscopist looks for any and all visible lesions, particularly polyps and cancers. Small adenomas can be completely excised with the colonoscope.

Colon adenoma—Precancer of colon and rectum that can sometimes progress to invasive cancer (adenocarcinoma of colon).

Colorectal—Refers to the colon plus the rectum. Surgeons insist on separating colon cancers from rectal cancers, because the way these two cancers are treated surgically is quite different.

Comorbidity bias—Cancer survival time is the period of time that patients live following a diagnosis of cancer. Whether a patient dies of cancer or heart disease or motor vehicle accident is immaterial. People with life-shortening comorbid conditions at the time of their cancer diagnosis will have a shortened survival time on this basis alone. People who are in generally good health when they are diagnosed with cancer, tend to live longer than people with comorbid conditions. Depending on the design of the clinical trial, comorbid conditions can bias the survival results.

Confounder bias—Unanticipated or ignored factors that alter an outcome measurement. Designing clinical trials that account for confounder influences is impossible, because confounders are unanticipated. Confounders are the statistical byproducts of the "Law of Unanticipated Consequences," which simply states that there will always be unanticipated consequences. Here's an example: Statins are widely used drugs that reduce the blood levels of cholesterol and various other blood lipids. To the best of my knowledge, nobody expected that the use of statins would have any effect on the incidence or mortality of cancer. In a recent study involving nearly a half-million male patients conducted between 1998 and 2004, statin use exceeding six months was linked to a significant lung cancer risk reduction of 55%. Participants who took a statin longer than four years had a 77% reduction in lung cancer risk (165). Let us suppose that the report is accurate and that we can eliminate deaths from lung cancer by 77% just by prescribing statins to everyone at risk for lung cancer. In the United States, about 90,000 men and 70,000 women will die of lung cancer this year (166). If this were reduced by 77%, we would prevent cancer deaths in 123,000 people. If there were unanticipated influences of statin use among some clinical trial participants, who take statins for reasons unrelated to the trial design, then this would be an example of confounder bias.

Congenital—Present at birth

Congenital tumors—Congenital tumors are neoplasms present at the time of birth. They are extremely rare. In an audit of 17,417 perinatal autopsies (autopsies performed soon after birth) conducted at the Royal Women's Hospital, Melbourne, Victoria, Australia, 46 congenital tumors were encountered: 24 teratomas, 8 vascular tumors, 6 neuroblastomas, 3 rhabdomyomas, 2 mesoblastic nephromas, 1 thyroid adenoma, 1 hepatic adenoma, and 1 cerebellar medulloblastoma were found (167).

Connective tissues—Structural soft tissues, including fibrous tissue, cartilage, adipose tissue, and vessels. The term sometimes includes muscle and hard tissue (bone).

Cyclosporine—Immunosuppressive agent used to treat patients who have received organ transplants. Treatment with cyclosporine reduces the likelihood that the body will reject the transplanted organ.

Cytogenetic—Related to chromosomal morphology. During mitosis, chromosomes elongate. Stained, elongated chromosomes have characteristic bands, corresponding to areas of chromosomal compression. By examining banded chromosomes under a microscope, a trained cytogeneticist can detect changes in chromosome number, gains and losses of whole chromosomes, translocations, duplications, losses, and complex rearrangements of large or small segments of chromosomes.

Cytologic—Refers to properties of individual cells, especially morphologic features viewed under a microscope by a trained cytologist or pathologist. Cytologic cell specimens are obtained from body fluids (urine, pleural fluid, peritoneal fluid, joint fluid), smears (Papanicolaou smears of cervix, scrapings from the mouth or esophagus), endoscopic brushings (of bronchus or pancreatic ducts), and needle aspirations of palpable masses or internal masses that have been located through radiologic imaging procedures. A total of over fifty million human cytology specimens are examined each year in U.S. cytology laboratories.

Desmoplasia—Reactive growth of nonneoplastic connective tissue adjacent to invading cancer cells.

Ductal carcinoma—Cancer arising from the epithelium of a duct. Ductal carcinomas are subtypes of adenocarcinomas (glandular carcinomas). The most common ductal carcinomas arise from ducts of the breast (ductal carcinoma of breast), pancreas (pancreatic adenocarcinoma), extrahepatic ducts (choledochal carcinoma), intrahepatic ductules (cholangiocarcinoma), and salivary glands (ductal carcinoma of salivary glands). Tumors arising from the bronchus are not considered to be ductal tumors, but they could be grouped as such, as the bronchus and bronchioles are specialized types of ducts that lead to the specialized airway glands of the lung (pulmonary alveoli).

Ductal intraepithelial neoplasia (DIN)—A generic term for precancers that arise in the ductal epithelium. Current usage of "DIN" is confined to the precancer for infiltrating ductal carcinoma of breast, the most common form of breast cancer.

Dysmyelopoiesis—Myelopoiesis is the production of normal blood cell lineages (red cells, white cells and platelets) from myeloid precursors. Normal myelopoiesis occurs in the bone marrow. In dysmyelopoiesis, abnormal cells are formed, often with functional impairment and shortened cell life span.

Dysplasia—Atypia that characterizes the cells in precancers.

Dysplastic nevus—Nevus with a set of characteristic morphologic changes that may precede the development of melanoma. Dysplastic nevi are present in kindred with an inherited predisposition to develop melanomas. Dysplastic nevi are also observed sporadically. An important goal of "mole screens" performed by dermatologists is to find dysplastic nevi and remove them before they can progress to become invasive melanomas.

Early cancer—"Early" cancers are not well-defined, and the term means different things to different people. Many people use it to mean a small cancer that has not had the chance to grow to a clinically detectable size. Other people include the precancers among the early cancers. As used in this book, an early cancer is any cancer that is detected earlier than usual through some interventional action, such as a screening test. With all these definitions, the term *early cancer* includes lesions that may have been around for years before detection.

Early detection—In the context of this book, early detection covers precancer detection and the detection of cancers before they have metastasized. The adjective *early* is somewhat deceptive, because the process of carcinogenesis can extend over years (sometimes decades) before metastasis occurs.

Ectocervix—Squamous-lined outer zone of the cervix. See *Endocervix*.

Electron microscopic—System that sends focused electrons (rather than focused light) through a specimen, to obtain a photograph. These focused electrons have a smaller wavelength than the wavelength of visible light and can resolve smaller objects. As a generalization, if you want to see objects measured in microns (millionths of a meter), use a light microscope. If you want to see objects measured in nanometers (billionths of a meter), use an electron microscope. For example, most viruses are 5 to 300 nanometers and can only be visualized under an electron microscope. Most animal cells range from 10 to 50 microns in diameter and can be well-visualized in a light microscope.

Embryonal carcinoma of testis—Malignant tumor derived from male germ cells (primitive cells that would normally produce differentiated spermatocytes).

Embryonic rest—Embryonic tissue that would normally differentiate to produce a fully developed tissue in the mature organism but which persists for some indefinite time. Embryonic rests are usually biologically inconsequential. In rare cases, neoplasms can originate from embryonic rests.

Endocervix—Glandular-lined inner zone of the uterine cervix. The cervix is partitioned into an inner endocervix (lined by glands) and an outer ectocervix (lined by squamous epithelium), separated by a transformation zone (squamocolumnar junction). See *Ectocervix*.

Endocrine—Endocrine cells produce substances that are carried in the bloodstream and exert a physiological response in receptor cells located in distant anatomic locations. Examples of endocrine organs are thyroid, parathyroid, pituitary, and adrenal glands.

Endometrial intraepithelial neoplasia (EIN)—Precancer arising from the epithelial lining of the uterus (the endometrium) (168).

Enteric endocrine—Any cell that secretes a substance that is carried in the bloodstream and produces a biological effect on a remote cell is an endocrine cell. Most endocrine cells are present in endocrine glands, specialized tissues composed of large numbers of endocrine cells (e.g., thyroid, adrenal, and parathyroid). The gut contains many endocrine cells dispersed throughout the gastrointestinal tract and scattered in the mucosal lining among mucus-producing enterocytes.

Enterocytes—Enteric cells, cells that line the alimentary tract. There are many different types of enterocytes, each specialized to perform a function that aids digestion. Absorptive cells and mucus-filled (goblet) cells are the most common enterocytes in the intestine. Most gastrointestinal cancers and precancers morphologically resemble absorptive cells or goblet cells.

Epidermis—Outer layer of skin, composed of multilayered squamous epithelial cells. The bottom (basal) layer of epithelial cells (cells closest to the underlying dermis) is the generative layer and contains a population of dividing cells that give rise to the nondividing cells that rise through the epidermis (over a period of weeks). As nondividing cells rise through the epidermis, they change their shape, structure, and chemical content. The top layer consists of flat, anucleate cells filled with keratin. These cells eventually slough into the air. Most house dust is sloughed, keratinized cells.

Epigenome—Collection of all the chemical modifications in the base molecules that compose the nuclear DNA sequence known as the genome, as well as other chromosomal constituents (such as histones and nonhistone proteins). The process of differentiation in normal cells involves inherited epigenetic modifications of chromosomes, including *DNA* methylation changes, and modifications to proteins that inhabit chromosomal material. The resulting differences among the many distinctive types of cells in the human body, all of which share an identical genome, is determined by the epigenome. The most studied epigenomic modification is DNA methylation.

Epithelial cell—Cell that is part of an epithelium. An epithelium is a collection of tightly fitting polygonal cells (epithelial cells) that cover a surface or form an organ composed of glands. Examples of epithelial structures are the epidermis (composed of squamous epithelial cells), the liver (composed of epithelial hepatocytes), the thyroid (composed of follicles lined by thyroid epithelial cells), and breast (composed of ducts and glands lined by epithelial cells). Examples of nonepithelial cells are connective tissue cells, muscle cells, bone cells, brain cells, hematopoietic cells, and lymphoid cells.

Epithelium—Tissue formed by epithelial cells. See *Epithelial cell.*

Euploid—Having the normal number of chromosomes for a somatic (body, nongerm) cell. In humans, the euploid chromosome number is 46. See *Aneuploid.*

Excision—Short for excisional biopsy. A biopsy intended to remove the entire lesion (usually, a primary neoplasm).

Fine needle aspirate (FNA) or *fine needle aspiration biopsy (FNAB)*—Cells from any normal or abnormal tissue in the body can be sampled by a needle inserted into the tissue. Cytologic samples obtained by this method are called fine needle aspiration biopsies, FNAs, or FNABs.

Fixatives—See *Formalin.*

Food and Drug Administration (FDA)—The U.S. FDA has the responsibility to ensure the safety of food and drugs. The FDA is also expected to promote innovation in the pharmaceutical field.

Formalin—Most commonly used fixative agent in histology laboratories. A fixative denatures macromolecules in cells, so that no biological activity can proceed. Formalin is a crosslinking fixative that denatures macromolecules by bonding molecules to one another. After fixation, a tissue is said to be "fixed." Fixed tissue, stored in fixative, is preserved indefinitely and can be sampled at any later date for histologic processing and microscopic examination.

Frozen tissue—Tissue fixation preserves tissue from autolytic activity (degeneration due to released enzymes) and from bacterial growth. Fixation also destroys the activity of enzymes and the structure of macromolecules that characterize cells of the tissue or that help elucidate the physiological status of cells at the time that the tissue was excised. To preserve the activity and structure of cell molecules while avoiding cellular changes that occur after excision, tissues can be fast-frozen and either used immediately or kept frozen and used at a later date.

Gardasil—Commercial name (Merck pharmaceuticals) for a vaccine active against four of the most common forms of carcinogenic human papillomaviruses.

Gastric dysplasia—Precancerous atypia of the stomach epithelium. Gastric dysplasia can follow intestinal metaplasia of the stomach epithelium.

Gastroesophageal reflux disease (GERD)—Backflow of stomach contents into the esophagus. GERD can cause esophageal erosion and ulceration and Barrett esophagus. Rarely, GERD can cause esophageal dysplasia and even esophageal carcinoma.

Gene expression arrays—Also known as gene chips. Thousands of small samples of DNA, arranged onto a support structure (usually, a glass slide). Matching DNA prepared from tissue samples is annealed onto the array. The measurement of annealing molecules provides an indication of the genes expressed by the cells in the tissue sample.

Gene mutation—Alteration in the cellular DNA sequence that is carried over into the DNA sequence of daughter cells (i.e., a heritable DNA alteration).

Genetic instability—Tendency to develop changes in cellular DNA (e.g., mutations, translocations, duplications, deletions, and so on) over time. Genetic instability seems to be an intrinsic property of all cancers and is responsible for cancer progression (accumulation of cellular alterations over time) and cancer heterogeneity (development of distinctive subclones within the cells of a tumor).

Genomic—The genetic material for an organism. Sometimes genome refers exclusively to the DNA sequence for an organism: this definition would restrict genome to a haploid sequence. Sometimes genome refers to all the genetic material characterizing an organism, including nuclear DNA, mitochondrial DNA, and the complement of expressed RNA.

Genotype—Characteristic genetic sequence for an organism. Every organism on earth has a unique genotype, with the possible exception of identical twins and some organisms that replicate asexually. See *Phenotype.*

Germline mutation—Mutation present immediately after fertilization and passed to every cell in the body of the developing organism.

Glass slide—To a pathologist, the basic diagnostic specimen is the glass slide. Surgically excised samples of tumors and other diseased tissues are cut into very thin sections, about 4 to 6 microns (millionths of a meter) in thickness, stained with dyes to bring out cellular detail, and mounted on glass slides. The slides are viewed on a microscope. The pathologist renders a diagnosis based largely on recognizing cellular alterations that characterize specific diseases.

Glycolysis—Metabolic pathway that generates energy (in the form of ATP) from glucose rather than from oxygen (aerobic respiration). Tumors that often grow in relatively anoxic conditions depend heavily on glycolytic metabolism.

Grade—Determination, made by a pathologist of the degree of atypia in the cells of a cancer. In general, cancers with a high degree of atypia have a worse prognosis than cancers with a low degree of atypia.

Hairy cell leukemia—Type of lymphoma in which the tumor cells have a distinctive morphology characterized by villi extending from cell membranes, simulating hairs. Tumor cells are present in the spleen, blood, and bone marrow. Long-term remissions are often achieved with new chemotherapeutic protocols.

Helicobacter pylori—Bacteria that grow in a thin layer overlying stomach epithelium. *H. pylori* produces acute and chronic gastritis (inflammation of the stomach), gastric ulcers (erosions of the stomach epithelium, producing bleeding), and contributes to the development of gastric adenocarcinoma and lymphomas of the stomach. *H. pylori* can be successfully treated with antibiotics.

Hematoxylin and eosin (H and E)—Combination of two stains commonly used to stain tissue on glass slides. The stained, mounted tissue sections are evaluated by pathologists.

Histology—Study of tissues, viewed with a microscope.

Hodgkin lymphoma—Lymphoma characterized by large, atypical binucleate cells (so-called Reed-Sternberg cells) and by a mixture of different types of lymphocytes. Many cases of Hodgkin lymphoma can be cured. Epstein-Barr virus (EBV), a known oncogenic virus, has been linked to Hodgkin lymphoma.

Human papillomavirus (HPV)—There are many strains of human papillomavirus. Certain strains of HPV cause common warts, laryngeal nodules, benign condylomas (warts) in the anogenital area, and cervical squamous cell carcinoma.

Hypercellularity—Greater-than-normal density of cells at a focus.

Hyperplasia—Overgrowth of normal tissue.

Hypoxic—Characterized by lower than normal levels of oxygen. Most cancers are relatively hypoxic and employ metabolic pathways that allow them to survive in hypoxic conditions.

Immune status—Important for cancers of viral origin. Immune system suppression is sometimes produced medically (as in transplantation procedures) and is sometimes caused by diseases (as in AIDS). Immunosuppressed individuals are prone to many diseases caused by endogenous infectious agents. Endogenous (commensal) infectious agents are organisms that live in our bodies throughout most of our lives without causing clinical disease, or they are organisms that create a mild form of disease that does not normally lead to death. These same organisms may, when the immune system is suppressed, break out as fulminant, life-threatening infections. Among viruses that lurk in humans are many tumor viruses. Shortly after a patient is immunosuppressed, sometimes in just a few months, the unconstrained proliferation of endogenous oncoviruses may produce malignancies, including Kaposi sarcoma (a type of angiosarcoma), primary central nervous system lymphoma, Hodgkin lymphoma, cervical squamous cell carcinoma, and primary effusion lymphoma.

Immunogenic—Capable of producing an immune response.

Immunosuppression—Condition in which the normal immune responses of the body are muted or eliminated because of a drug (e.g., cyclosporine, cortisone), a physical effect (e.g., radiation) an infection (HIV), or an inherited disease (severe combined immunodeficiency).

Immunotherapy—In the context of cancer, immunotherapy is a cancer treatment that boosts the immune system in hopes that a heightened immune response can destroy cancer cells.

In situ—Latin: in place. Indicates that a neoplasm is noninvasive.

Incidence—Number of new cases occurring in a unit of time (usually a year). Usually expressed as a rate per 100,000 population. See *Age-adjusted incidence, Prevalence.*

Initiation—First step in carcinogenesis that occurs when a population of cells is exposed to a carcinogen.

Intestinal metaplasia—Foci in the esophagus or stomach where normal esophageal or gastric tissue histology is replaced with tissue that more closely resembles the histologic appearance of the intestines (i.e., glands lined by mucus-producing goblet cells and absorptive enterocytes). Seen in Barrett esophagus and chronic gastritis.

Intraductal hyperplasia (IDH)—Although the term can be applied to epithelium within any duct, the use of this term is commonly restricted to the breast. IDH precedes atypical ductal hyperplasia.

Intraepithelial neoplasia (IEN)—Class of precancers that derive from an epithelial mucosa or epithelial lining, delimited by an anatomically distinct basement membrane. Often, the term *intraepithelial neoplasia* is used interchangeably with the term *precancer*. This is simply wrong, because the term *precancer* includes many types of lesions that are not intraepithelial (List G.1). Throughout this book, we use the term *precancer* to include the intraepithelial neoplasias as well as other identifiable lesions that precede the development of malignant neoplasms.

Invasion—Intrusion of cancer cells into normal tissues.

Keratin—Protein produced normally in large amounts by squamous cells that line the skin. As squamous cells rise through the epidermis (a process that typically takes four weeks), their cytoplasm becomes filled with keratin. At upper levels of the squamous epithelium, cells are completely filled with keratin protein. The cell dies, leaving a keratin husk the size and shape of the cell. This layer of husks is the keratinized top layer of skin. These keratin husks slough into the air, to be replaced by the dying squamous cells ascending from the next lower level of the epidermis.

List G.1 Examples of precancers that are not IENs.

Myelodysplasias and myeloproliferative disorders

Lymphoproliferative lesions

Nodular epithelial lesions for which there is no surrounding basement membrane, including preneoplastic nodules of liver, cortical adenomas of kidney, and thyroid C-cell proliferative lesions that precede medullary carcinoma of the thyroid

Dysplastic nevi that may have both intraepithelial and dermal components

Noninvasive neoplasms of nonepithelial origin, such as intratubular germ cell neoplasms

Identifiable precursors for sarcomas

Keratoacanthoma—Squamous cell neoplasm that grows rapidly, over a period of weeks to months, then regresses. In rare instances, keratoacanthomas progress to squamous cell carcinoma.

Lamina propria—Connective tissue that underlies a mucosal (epithelial) surface.

Latency—After initiation, years may pass before a precancer or cancer emerges. This lag in visible biological progression is called latency, or the latent period.

Lesion—Focal area of tissue involved by a pathologic process (degenerative, inflammatory, toxic, or neoplastic) that is characteristic of a particular disease. In the context of this book, a lesion is a focus of neoplasia (precancer or cancer).

Leukemia—Cancer of blood cells.

Liver cancer—Most liver cancers arise from hepatocytes, the cells that perform most metabolic processes of the body. Cancer arising from hepatocytes is hepatocellular carcinoma. Other cancers of the liver include cholangiocarcinoma (cancer in liver ductules that carry bile to the extrahepatic duct system), angiosarcoma, and hepatoblastoma (rare cancer of childhood). Hepatocellular carcinoma often arises from liver cirrhosis, a disease in which there is massive liver fibrosis (scarring), with death and regeneration of enlarged, architecturally distorted liver acini (glands). The number of deaths in the United States from liver cancer has risen steadily over the past half century (Figure G.2). Worldwide, hepatocellular carcinoma is the third leading cause of cancer deaths.

Lymphoma—Cancer of lymphocytes. Currently, all lymphomas are considered malignant, and lymphomas have no generally recognized precancers.

Malignancy—Malignant tumor; synonymous with cancer. A tumor that progressively grows, invades, and, if left untreated, kills.

Marker—As used in this book, a marker is a cellular abnormality that identifies or characterizes a precancer or cancer. For example, the Philadelphia chromosome is a cytogenetic marker for chronic myelogenous leukemia.

Marketing bias—In a fascinating meta-analysis, Yank and coworkers wanted to know whether the results of clinical trials conducted with financial ties to a drug company were biased toward favorable results (20). They reviewed the literature on clinical trials for antihypertensive agents and found that ties to a drug company did not bias the results. They found, however, that financial ties to a drug company are associated with favorable conclusions. This suggests that regardless of the results of a trial, the conclusions published by the investigators were more likely to be favorable, if the trial were financed by a drug company. This should not be surprising. Two scientists can look at the same results and draw entirely different conclusions. Marketing bias occurs when the popularity of a drug is determined primarily by the effectiveness of its promotion.

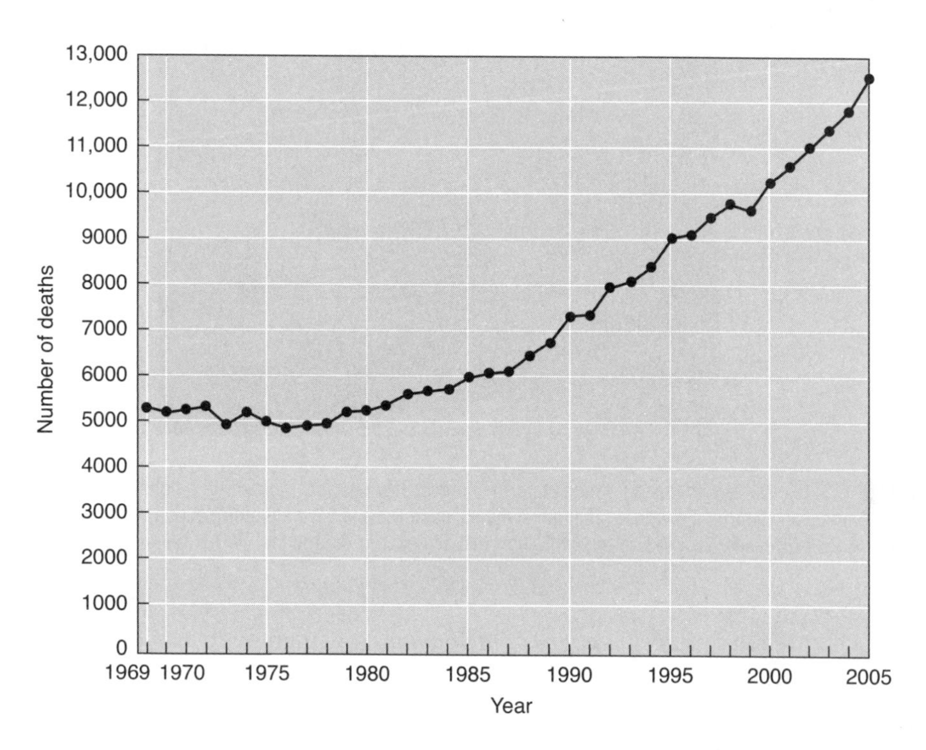

FIGURE G.2 Deaths from hepatocellular carcinoma in the United States. The Y-axis represents the total number of deaths from hepatocellular carcinoma occurring in the United States in a given year. Data provided by the NCI's SEER program (http://seer.cancer.gov/cangues/mortality.html).

Mastocytosis—Proliferation of mast cells (tissue cells derived from a bone marrow precursor). Most often, the term refers to a neoplastic proliferation of mast cells, not simply a reactive and transient increase in mast cells.

Measurement bias—The accuracy of response measurements may be poor. For example, some of the newest anticancer drugs have low toxicity for cancer cells. Current and future drugs may act by decreasing the growth rate of a tumor or reducing the likelihood that the tumor will metastasize. These effects may not shrink or eradicate a tumor, but they may increase the length and quality of postdiagnosis survival without changing the long-term mortality of the cancer. In these cases, the measurements of tumor response and patient survival should provide useful information related to the value of the treatment without producing an unjustified expectation of cure.

Melanoma—Malignant skin tumor derived from the melanin-producing cells of the lower epidermis. Most melanomas are pigmented (Figure G.3). Amelanotic melanomas are melanomas that do not produce melanin. The single word *melanoma* is always assumed

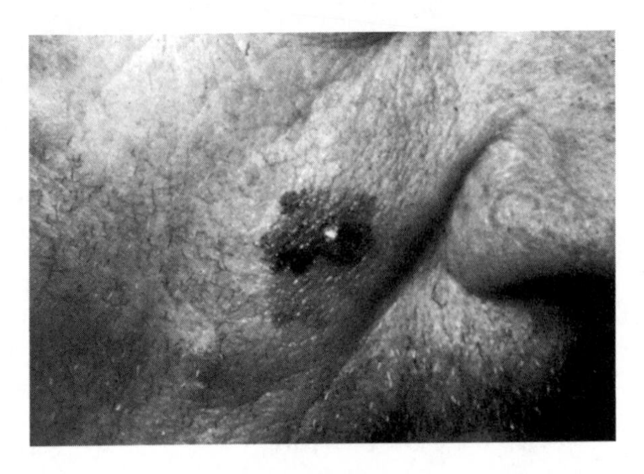

FIGURE G.3 Clinical appearance of a malignant melanoma. The lesion is mottled with light and dark areas, raised and flat areas, and an irregular border. Benign nevi are symmetric and have a uniform appearance. From *An Introduction to Human Disease*, Seventh Edition. Photo courtesy of Leonard V. Crowley, M.D., Century College.

to be a shortened form for *malignant melanoma.* A benign neoplasm of melanocytes is called a nevus.

Mesoderm—In embryonic development, mesoderm is a primitive layer of loose tissue that resides between the layers of endoderm (inner layer) and ectoderm (outer layer). By weight, the mesoderm gives rise to most of the adult human body, providing connective tissue, muscle, and bone, myelopoietic (blood cell) system, lymphopoietic (lymph node) system, and several organs (including uterus and kidneys).

Meta-analysis—A study that combines and analyzes data from multiple other studies.

Metabolic pathway—According to traditional thinking, a metabolic pathway was a sequence of biochemical reactions involving a specific set of enzymes and substrates that produced a chemical product. The classic pathway was the Krebs cycle. It was common for students to calculate the output of the cycle (in moles of ATP), based on stoichiometric equations employing known amounts of substrate. Today, in the field of cancer research, metabolic pathways operative in tumors usually involve linked actions among receptors, activators, enzymes, and sites in macromolecules. Most pathways produce specific cellular activities; not measurable chemical products. Pathways may not always occur within a specific organelle. Pathways may interact with other pathways, and their directions and biological consequences may be complex and variable among different cell types or in different physiological states within one cell type. Still, the term *metabolic pathway* is a convenient conceptual device, to organize classes of molecules that interact with a generally defined set of partner molecules that produce a consistent range of biological actions.

Metaplasia—Replacement of normal cells in a tissue with cells that are normally present in a different tissue.

Metastasis—Spread of a cancer to distant tissues, through seeding via blood vessels or lymphatic vessels.

Monoclonal gammopathy of undetermined significance (MGUS)—Monoclonal increase in plasma cells, with a resultant spike in circulating levels of the specific antibody synthesized by the clonal plasma cells. MGUS is actually a common condition in the older population, and may occur in about 1% of the population over 70 years of age. Progression of MGUS to multiple myeloma is infrequent, with a conversion of about 1% to 2% per year (70). Because MGUS occurs in an older population, the chance of MGUS progressing to myeloma within the lifespan of the patient is quite low. MGUS is an example of a precancer that is not an intraepithelial neoplasm. See *Intraepithelial neoplasia (IEN).*

Mitosis—Stage in cell division where chromosomes migrate to the two daughter cells (Figure G.4).

Mole—Synonym for *nevus.*

Molecular signatures—Characteristic gene expression profiles for a collection of tissue samples (usually cancers).

Molecular targeted drugs—Chemotherapy that selectively targets the activity of a few molecular species within the cell. For example, Avastin (Bevacizumab) binds to and inhibits the biologic activity of human vascular endothelial growth factor (VEGF); Iressa blocks EGFR; Gleevec (imatinib mesylate) inhibits ABL, PDGFR, and KIT kinases; SU11248 inhibits c-KIT, VEGFR, and PDGFR. Because these agents target a narrow range of cellular pathways that are elevated in cancers, they produce very little systemic toxicity. Most targeted chemotherapies are also called nontoxic chemotherapies.

Morbidity—Unhealthy condition or disease.

Mortality—Latin: *mortis*, of death. Medical euphemism for death.

Mucosa—Surface epithelial layer plus the loose underlying connective tissue (lamina propria). The mucosa ends where the submucosa begins, at the level of the muscularis mucosae.

Mucosa-associated lymphoid tissue lymphoma (MALToma)—Most of the gastrointestinal tract and salivary glands are lined by lymphoid tissue that lacks the typical architecture of lymph nodes. Basically, the lymphoid tissue that lines the gut mucosa consists of aggregates of lymphocytes with some lymphoid follicles. Most MALT cells are specialized B lymphocytes (monocytoid B lymphocytes) that share many features with marginal zone B lymphocytes present in lymph nodes. Tumors that arise from MALT tissues are called MALTomas. MALTomas may occur in the stomach after chronic infection with *Helicobacter pylori.* MALTomas sometimes regress when the *Helicobacter* infection is treated.

Multiple endocrine neoplasia (MEN)—One of several inherited syndromes for which tumors may arise from several different endocrine glands, as well as several nonendocrine tissues.

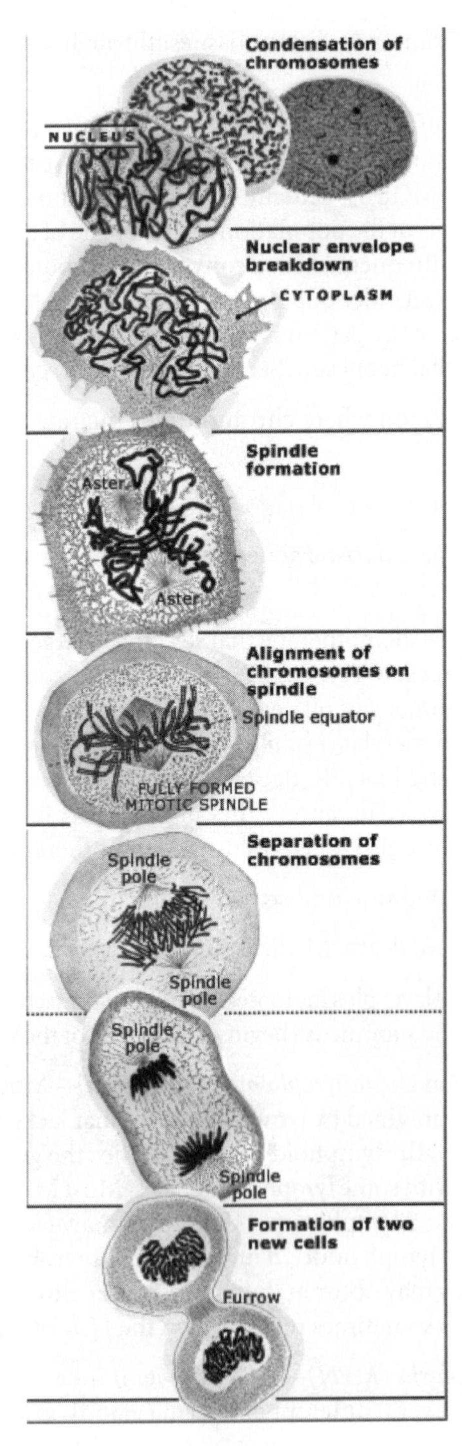

FIGURE G.4 Mitosis. Reproduced from Flemming, W., 1879. *Archiv fur Mikroskopische Anatomie.* Berlin: Cohen.

Multiple myeloma—Malignant neoplasm of neoplastic plasma cells. Plasma cells are blood cells that produce immunoglobulins. Blood contains many unique immunoglobulin molecules, each present in small concentration, produced by many different normal plasma cells. Most multiple myelomas are associated with a single immunoglobulin molecule, present in blood in high concentration, produced by all the tumor cells, and specific for the tumor. This is called an immunoglobulin spike. The spike is evidence of the clonal nature of multiple myeloma. The tumor is called multiple myeloma because it usually presents with multiple tumor nodules in different locations of the bone marrow and elsewhere.

Mutation—Alteration of cellular DNA that passes to a daughter cell. When a mutation is produced in a germ cell of an organism and the germ cell is fertilized, then there is a chance that the mutation will be carried into every cell of the resulting organism. A mutagen is an agent that can produce a mutation.

Mycosis fungoides—Lymphoma of T lymphocytes that occurs in the skin. When mycosis fungoides cells are present in the blood, the constellation of clinical findings is often called Sezary syndrome. Despite its name, mycosis fungoides is not caused by a fungus.

National Cancer Institute (NCI)—Largest research institute of the U.S. National Institutes of Health. The NCI has an annual budget of about $5 billion, which is spent on research and clinical trials conducted at its campus in Bethesda, Maryland, and at universities and medical centers throughout the United States and in partner countries.

Neoplasm—Greek: *neos*, new; *plasso*, shape. General term that includes benign tumors, precancers, and cancers.

Nephroblastoma—Also known as Wilms tumor. Rare kidney tumor that occurs almost exclusively in children. Histologically it is composed of primitive cells of renal origin.

Nephrogenic rests—Remnants of fetal kidney that sometimes persist after birth. In some cases, Wilms tumors arise from nephrogenic rests.

Neuroblastoma—Rare tumor of the peripheral nervous system (the nervous system not included in the central nervous system of the brain and spinal cord) that develops from precursor cells of neurons.

Nevus—Benign, small papule on the skin, almost always pigmented (brown or black). The medical significance of nevi is that they may rarely become cancerous. The cancer that can arise from a nevus is a malignant melanoma.

Non-Hodgkin lymphoma—All lymphomas other than Hodgkin lymphoma. See *Hodgkin lymphoma.*

Nontoxic chemotherapies—Chemotherapy that selectively inhibits the growth of cancer cells without producing much systemic toxicity. See *Molecular targeted drugs.*

Nuclear atypia—In normal tissues, the nuclei of any cell type have a uniform morphology. Each nucleus has roughly the same size, shape, and texture (distribution of chromatin). In precancers and cancers, nuclei are almost always characterized by aberrations from normal. Precancerous and cancerous nuclei are almost always larger than normal nuclei of the same cell type. Simply knowing the nuclear sizes of a population of cells is often sufficient to make a cytologic diagnosis of precancer or cancer. Precancer and cancer nuclei are characterized by nonuniformity. The size of a nucleus varies from cell to cell within the lesion. The same is true for every other feature of the precancerous and cancerous nucleus. Some nuclei are round; others have small indentations in the nuclear edge; and others are smooth but show concavity. The area of concavity may change from cell to cell, sometimes facing into the center of the cell and sometimes facing the nearest cytoplasmic border. The nuclear texture varies greatly from focus to focus within the nucleus and between one nucleus and the next. Areas of chromatin clumping (focal hyperchromasia) alternate with areas of nuclear hypochromasia (lack of color, parachromatin clearing). The coarse texture of chromatin in precancerous and cancerous cells is different from the fine chromatin texture of normal cells.

Nucleolus—Area of the nucleus where messenger RNA (mRNA) is synthesized.

Nucleus—The membrane-bound compartment characteristic of eukaryotic organisms that contains genetic material (DNA), epigenetic modifications to DNA, epigenetic material (histones and nonhistone proteins), and a matrix of fibers through which a variety of molecules diffuse and interact.

Oncogene—Normal genes that, when altered, can contribute to the carcinogenic process. Many different cancers of humans have been shown to contain oncogenes that are characteristic for the tumor.

Oncologist—Physician who specializes in treating patients who have cancer.

Ovarian intraepithelial neoplasia (OIN)—Putative precursor for the common forms of ovarian cancer that arise from ovarian surface epithelium. We know very little about OIN at this time.

Pancreatic intraepithelial neoplasia (PanIN)—Pancreatic cancer is a highly aggressive malignancy and is the fourth most common cancer killer in the United States. We know almost nothing about the causes of pancreatic cancer. Most pancreatic cancer arises from ductal lining cells (that carry secretions from the pancreatic glands into the gut). Pancreatic precancer, PanIN, is characterized by focal, morphologic atypia of proliferating duct cells.

Pap smear—Short for Papanicolaou smear. A cytologic specimen obtained from the uterine cervix.

Papanicolaou—George N. Papanicolaou (1883—1962) is generally considered the father of clinical cytology. The principles of cytologic evaluation of cervical precancer were established in his book, *Diagnosis of Uterine Cancer by the Vaginal Smear* coauthored with Herbert Traut and published in 1943.

Pathogenesis—Biological events that lead to the expression of a disease. Carcinogenesis is a special instance of pathogenesis.

Pathology data sets—Pathologists boast that they produce 70% of the information in hospitals, that a pathology report is rendered on 70% of hospital patients, and that 70% of decisions made in a clinical setting are based on pathology data. These figures are somewhat fanciful, but nobody doubts that pathologists produce an enormous amount of very useful medical data, that are archived in hospital information systems. The kinds of data produced by pathologists are: clinical laboratory test results (e.g., quantitative blood tests), microbiology reports, blood usage reports, surgical pathology reports, cytology reports, and autopsy reports.

Persistence—A biological feature of cancer. Unless excised or treated with chemotherapeutic agents, cancer persists and grows indefinitely. Persistence is not a constant feature of precancers. Over time, some precancers regress (lose persistence). The mechanisms for persistence and regression are unknown to cancer scientists.

Phenotype—The totality of cell or tissue behavior expressed in the morphology and cellular physiology of a cell or tissue. Phenotype is determined by structural features, enzymatic pathways, and environmental influences. See *Genotype*.

Philadelphia chromosome—Almost all patients with chronic myelogenous leukemia have a characteristic translocation called the Philadelphia chromosome. Parts of chromosomes 9 and 22 translocate and form a fusion gene composed of a part of the breakpoint cluster region (BCR) gene and a part of the Abelson leukemia (ABL) gene (see Figure 5.3).

Phlebotomists—Medical technicians trained to collect blood samples from patients. Phlebotomy involves inserting a sterile needle into a vein that has been engorged by an elastic tourniquet. Blood is drawn into one or several sterile vacuum tubes. The process is usually conducted without harming the patient or the phlebotomist and yields a properly identified sample of adequate volume that is neither hemolyzed nor clotted.

Pleural fluid—All body cavities are lubricated with fluid. These spaces include the space between the lungs and the ribcage (the pleural cavity), the space between the intestines and the abdominal wall (the peritoneal cavity), and the space between adjacent bones (the joint cavities). Under normal circumstances, cavities are lubricated by a small amount of fluid secreted by the cells that line the cavities. Under pathological conditions (inflammation or malignancy), body cavities can fill up with fluid containing abnormal cells. Cytologists can examine samples of fluid taken from any body cavity and determine whether the fluid contains malignant cells.

Polyp—Protrusion of tissue. A colon polyp protrudes from the mucosal surface of the colon. There are two common types of colon polyp: hyperplastic polyp and adenomatous polyp. The hyperplastic polyp is a nonneoplastic focal area of exuberant growth of otherwise normal intestinal mucosa. Hyperplastic polyps have no clinical significance. Adenomatous polyps are benign, precancerous growths of colonic epithelial cells. Grossly, adenomatous polyps are either stalked or sessile. Stalked polyps are separated from the mucosa by a thin cord, composed of fibrovascular tissue lined by nonneoplastic colonic mucosa. Stalked adenomatous polyps are easy to excise by cutting the stalk. Sessile adenomatous polyps have no stalk and merge with the normal mucosa. They are somewhat more difficult to excise. Adenomatous polyps can range greatly in size from just a few millimeters to five or more more centimeters. The larger the adenomatous polyp, the greater the likelihood that it contains a focus of invasive carcinoma.

Population and demographic biases—For many different reasons, some populations accrue more easily into clinical trials than other groups. Children and pregnant women are sometimes excluded from clinical trials. In such cases, drug effectiveness and safety cannot extend to these under-represented groups. Third-party payers may refuse to cover the costs of drugs for pregnant women and children because of the lack of trial evidence ensuring drug safety and effectiveness in these groups. Demographic bias is a variant of population bias. In most clinical trials in the United States, patients are assigned into broad demographic groups, and the patients are often allowed to assign themselves into groups based on ethnicity. Treatment response rates may differ from group to group and in uncounted subgroups (e.g., Japanese patients may have a different response from Chinese patients, though both may be grouped as Asians). In clinical trials, groupings are usually based on stage of disease and age. Seldom do clinical trials stratify patients by income. Nonetheless, socioeconomic status greatly influences cancer survival (23). Groups that contain many economically disadvantaged patients may have a different outcome than similar demographic groups in which the members are financially well-off. Population bias effects every population that is not included in study populations.

Precancerous condition—Disease that predisposes a person to develop a precancer. Precancerous conditions can be inherited or acquired. Examples of inherited precancerous conditions are Milroy disease (hereditary lymphedema) and Fanconi syndrome. Examples of acquired precancerous conditions are AIDS, aplastic anemia, and liver cirrhosis.

Preleukemia—Precancerous lesion for a leukemia. The term *preleukemia* is seldom used. Preleukemias comprise the myelodysplastic disorders and the myeloproliferative disorders. See Chapter 5.

Prelymphoma—Although there is a bewildering variety of described lymphomas, there are no generally accepted lymphoma precancer lesions. We can infer, however, that they must exist. MALTomas, which arise in the stomach, often in chronic *Helicobactor pylori* infection, may regress when the infection is treated with antibiotics. The tumor regression is indicative of precancer, not cancer. In addition, nearly one-fourth of low-grade lymphomas regress if left untreated (90). The high regression rate of low-grade lymphomas suggests that some of the lesions that are currently diagnosed as lymphomas would be more accurately diagnosed as prelymphomas.

Prevalence—The number of people in a population, at a chosen moment in time, who have a disease. Chronic diseases can have low incidence and high prevalence. When a patient has a mild chronic disease diagnosed early in life, then the patient will be counted as a prevalent case through every year of a potentially long lifespan. An acute disease with high incidence may have low prevalence. If a fulminant illness lasts only a few days, the likelihood that any affected patient will be included in a count of prevalent cases is low. Treatments that extend lives without actually curing patients, will increase the prevalence of the disease. The burden of cancer on the healthcare system is strongly influenced by the prevalence of disease. Cancer treatments that extend survival, without achieving cures, inevitably increase cancer prevalence. When cancer researchers report survival results for clinical trials, predicting how the treatment might increase the prevalence of cancer in the population would be useful. See *Incidence.*

Primary tumors—Tumors located at the anatomic location where they originally developed.

Profile—In the context of this book, a profile or signature is the expression pattern of many different genes in a tissue, as measured on a gene chip. Researchers have found that the gene chip signature of a cancer has diagnostic relevance (169). The profile of a tumor can be used to find biological variants of the tumor or to predict the clinical behavior of a tumor (170, 171).

Progression—Acquisition, over time, of genetic alterations, subclonal heterogeneity, increasing cytologic atypia, continuous growth, and increasingly aggressive behavior in cancers. The progression phase begins at the moment when a precancer becomes a cancer. In most instances, progression continues relentlessly unless effective treatment is received.

Prostate cancer—Although several different forms of cancer may arise from the prostate gland, the most commonly occurring prostate cancer is adenocarcinoma. As used in this book, the term *prostate cancer* refers to adenocarcinoma of the prostate occurring in adult men.

Prostatectomy—Resection of the prostate gland.

Prostatic intraepithelial neoplasia (PIN)—The precancer for prostatic carcinoma.

Proteomic arrays—Several methods permit proteins found in fluids (e.g., blood) and tissues (e.g., cancer cells) to be separated (by molecular weight, charge, or both) and displayed as an array of many different protein species. The relative quantity of the different proteins produces a profile. Profiles may be used for screening, diagnosing, or predicting biological responses for many disease conditions.

Rare tumor—The definition of rare disease is included in the U.S. Orphan Drug Act of 1983 (172).

> *2). For purposes of paragraph (1), the term "rare disease or condition" means any disease or condition which (A) affects less than 200,000 persons in the United States, or (B) affects more than 200,000 in the United States and for which there is no reasonable expectation that the cost of developing and making available in the United States a drug for such disease or condition will be recovered from sales in the United States of such drug.*

With just a few exceptions, all cancers affect fewer than 200,000 people in the United States each year. So the official U.S. definition of a rare disease in not particularly helpful. The National Cancer Institute has defined a rare cancer as a cancer with under 40,000 new cases in the United States each year. This definition also includes all but a handful of cancer types and does not help distinguish a special set of tumors. In my opinion, a rare cancer is one that occurs with an incidence under 1 per 100,000 persons per year (i.e., under 3000 new cases in the United States each year). A rare tumor may be encountered once every few years in busy hospital. For any given rare tumor, the average oncologist might encounter a patient with that tumor once or never in a career.

Reabstraction bias—Trialists draw information related to the diagnosis, treatment, and outcome of patients by reviewing medical records. The quality of medical research often depends on the quality of medical records. When medical records are incomplete, incorrect, illegible, or otherwise uninterpretable, the results for an otherwise well-planned clinical trial can be disastrous. What do researchers do when they find that their medical records are inadequate? Often, they resort to reabstraction, a time-consuming, expensive, and occasionally futile undertaking. Reabstraction involves revisiting charts, visiting outpatient clinics and the private offices of medical doctors, reinterviewing patients and families, and a host of extraordinary efforts aimed at restoring credibility to clinical trial data. If one subset of patients has better-maintained records than another subset, then a bias can be introduced to the trial.

Reactive atypia—Nuclear morphology of normal cells displays a wide range of changes in response to toxic or physical damage that closely resembles the changes observed in cancer cells. The most prominent changes observed in regenerating normal cells are nuclear enlargement and nucleolar enlargement, both seen in cancer cells. In cells that have been exposed to a toxic agent (such as ionizing radiation), nuclear enlargement can be accompanied by variations in the shape of the nucleus, and variations in the degree of nuclear alteration from cell to cell. Trained cytologists can almost always distinguish the atypia of precancerous and cancerous cells from reactive atypia, when they have a technically good cytologic sample.

Regression—In the context of this book, regression is the cellular process that leads to the disappearance of a tumor. It typically involves the gradual reduction of cells in a population (such as a precancer) until the population has dwindled to zero. Regression occurs commonly in precancers. A precancer will either regress, stabilize (persist), or become a cancer.

Resection—Removal (cut out) in a surgical procedure.

Retinoblastoma—Tumor arising from primitive cells that produce the retinal lining epithelium in the eye.

Sarcoma—Malignancy arising from mesenchymal (connective) tissues, including fibrous tissue, adipose tissue, vascular tissue, cartilage, bone, and muscle.

Second trial bias—After a therapeutic trial, clinicians can determine the types of patients who are most likely to benefit from an intervention. For example, a trial of bone marrow transplantation for patients with metastatic carcinoma may indicate that patients over the age of 55 have a very poor response to transplantation. Older individuals with transplants may be more prone to die from the interventional procedure itself than from their cancer. On the second trial of the procedure, the clinicians will wisely exclude patients over some determined age (different from the inclusion criteria from the first trial). The second trial shows markedly improved survival compared to the first trial for those patients receiving bone marrow transplantation. An improved outcome in the second clinical trial can, thus, be achieved simply through better selection of subjects without any improvement in the treatment protocol.

Seminoma—Cancer arising from differentiated germ cells in the male. The equivalent tumor in females is dysgerminoma.

Small cell carcinoma—Highly malignant cancer composed of small cells with scant cytoplasm. Most small cell carcinomas arise from the lung. Small cell carcinomas of lung often metastasize widely while the primary lung cancer is still small. For this reason, surgical resection is seldom the best treatment for these tumors, which are more often treated with systemic chemotherapy. Because treatment for other types of lung cancer often benefits from surgery, oncologists conveniently divide all lung cancers into two categories: small cell carcinoma and non–small cell carcinoma.

Sphincter—A ring of muscle that closes or opens a passage from one anatomic structure to another. The normal position of a sphincter is constricted (closed). Sphincters relax to open. Examples of sphincters are the lower esophageal sphincter, the pyloric sphincter, the anal sphincter, and the sphincter in the iris of the eye.

Sporadic cancers—Tumors that occur without a demonstrated cause, apparently by chance.

Stage—Stage number (e.g., Stage 1, 2, 3, or 4) indicates the increasing extent of tumor spread.

Staging—Determination of the amount of spread of tumor at the time of diagnosis.

Statistical method bias—A statistician can look at a set of data, apply different statistical methods to the data, and arrive at different conclusions. In some cases, statisticians can draw opposite conclusions from the same set of data. Consequently, articles with opposite conclusions appear in the medical literature, permitting scientists to selectively cite those papers that support their own agendas (173).

Subclones—Clone that grows out of another clone. The subclone must be measurably different from the parent clone. The emergence of subclones within a tumor results in tumor heterogeneity.

Sulindac—Nonsteroidal antiinflammatory drug (NSAID) marketed in the United States by Merck & Co., Inc., under the name Clinoril.

Surrogate marker—Observation or measurement that substitutes for another observation or measurement. For example, an agent that induces a precancer is likely to be a carcinogen (something that induces a cancer). The precancer is a surrogate marker for an induced cancer in a bioassay test.

Surveillance Epidemiology and End Results (SEER)—Project within the U.S. NCI that collects and publishes various statistics on U.S. cancers. SEER data are generally considered the most authoritative source of data on cancers occurring in the United States.

Tamoxifen—Nonsteroidal estrogen antagonist used primarily to reduce the likelihood of cancer recurrence in women who have received breast cancer treatment.

Tissue microarrays (TMA)—First introduced in 1998, TMAs are collections of hundreds of tissue cores arrayed into a single paraffin histology block. Each TMA block can be sectioned and mounted onto glass slides, producing hundreds of nearly identical slides. TMAs permit investigators to use a single slide to conduct controlled studies on large cohorts of tissues, using a small amount of reagent. The source of tissue is only restricted by its availability in paraffin tissue blocks and ranges from cores of embedded cultured cells to tissues from any higher organism.

Translocation—In the context of cytogenetics, translocation is the movement of a section of a chromosome to some other chromosomal site.

Trialists—In a medical context, a trialist is a professional specially trained to initiate, design, and perform clinical trials.

Tumor—Latin: swelling. In this book, the term *tumor* is used as a synonym for neoplasm.

Tumor heterogeneity—Occurrence, in a single clonal tumor, of multiple subclones, each with a distinctive morphology or behavior.

Underreporting bias—It is human nature to celebrate success and bury failure. When a clinical trial produces negative results (fails to show improved survival), there may be little enthusiasm to publish the work. Sponsors of negative studies may be disinclined to rally the cancer research community and the public around their negative results. Dickerson and Rennie have written, "The fact that some trial results are never published would not be a problem, except that there is good evidence that the results from unpublished trials are systematically different from those of published trials" (31).

Virally induced—Some cancers are caused by viruses. Examples are Epstein-Barr virus (B-cell lymphomas, Burkitt lymphoma, nasopharyngeal cancer, Hodgkin lymphoma, and T-cell lymphomas); hepatitis B virus (hepatocellularcarcinoma); human papillomavirus types 5, 8, 14, 17, 20, and 47 (skin cancer); human papillomavirus types 16, 18, 31, 33, 35, 39, 45, 52, 56, and 58 (cervical cancer, anogenital cancer); human papillomavirus types 6 and 11 (verrucous carcinoma); human papillomavirus types 16, 18, 33, 57, and 73 (cancers of oral cavity, tongue, larynx, nasal cavity, and esophagus); HTLV-1 (adult T-cell leukemia); human herpesvirus 8 (Kaposi sarcoma); hepatitis C virus (hepatocellular carcinoma and low-grade lymphomas); JC, BK, and SV40-like polyoma viruses (tumors of brain, pancreatic islet tumors, and mesotheliomas); and human endogenous retrovirus HERV-K (seminomas and germ cell tumors).

War on cancer—Term used by President Richard Nixon in 1971 to characterize efforts to defeat cancer. The war on cancer was enacted by the National Cancer Act, which greatly strengthened the National Cancer Institute (the largest institute in the U.S. National Institutes of Health). The National Cancer Act became law on December 23, 1971. In 2008, the appropriation for cancer research, funded through the National Cancer Institute, was 4.8 billion dollars. This sum is a small fraction of the money spent treating the nearly 1.5 million (nonskin) cancers occurring in the United States each year and does not include cancer-related funding from other NIH institutes, from other government sources (such as the Department of Defense), from charitable organizations, and from private industry (such as the pharmaceutical industry). Many countries other than the United States expend great sums on cancer. Considering casualty rates (7,000,000 deaths per year worldwide), expenditures, and duration (38 years, so far), the war on cancer is the largest effort devoted to any individual disease and accounts for a loss of human life exceeding World War I and World War II.

Xeroderma pigmentosum (XP)—Rare inherited condition in which patients have reduced ability to repair certain types of DNA damage, including damage produced by ultraviolet light. Patients with XP develop skin cancer at an early age unless they strictly avoid exposure to sunlight.

Zymogen—Most enzymes are synthesized in an inactive form that is activated at the site where activity is needed. The precursor form of an enzyme is called a zymogen. Most digestive enzymes produced by the pancreas are released as zymogens that are activated in the intestines. If they were released in an active form, they would autodigest the pancreas.

References

1. Strategic Investments in Cancer Prevention, Early Detection, and Prediction. In *The Nation's Investment in Cancer Research. A Plan and Budget.* U.S. National Cancer Institute, 2006. http://plan2006.cancer.gov/prevention.shtml.
2. SEER. *Estimated new cancer cases and deaths for 2008.* http://seer.cancer.gov/csr/1975_2005/results_single/sect_01_table.01.pdf.
3. Baxter NN, Goldwasser MA, Paszat LF, Saskin R, Urbach DR, Rabeneck L. Association of colonoscopy and death from colorectal cancer. *Ann Intern Med* 150:1–8, 2009.
4. Lopez AD, Mathers CO, Ezzati M, Jamison DT, Murray CJ. Global and regional burden of disease and risk factors, 2001: Systematic analysis of population health data. *Lancet* 367:1747–1757, 2006. Two disease processes, cancer and cardiovascular diseases (including cerebrovascular disease) together accounted for 64.7% of all deaths occurring in the developed countries.
5. SEER. *Cancer Statistics Review 1975–2004.* National Cancer Institute. http://seer.cancer.gov/csr/1975_2004/results_merged/topic_lifetime_risk.pdf.
6. SEER. *Cancer Statistics Review 1975–2004.* National Cancer Institute. http://seer.cancer.gov/csr/1975_2004/results_merged/sect_01_overview.pdf, pages 30–31, Table I-4.
7. Hoffman FL. *The Mortality from Cancer throughout the World.* Newark: Prudential Press, 1915.
8. SEER. *Cancer Statistics Review 1975–2005.* http://seer.cancer.gov/csr/1975_2005/results_merged/topic_historical_mort_trends.pdf. Table I-2, 56-year trends in U.S. cancer death rates.
9. Hoyert DL, Heron MP, Murphy SL, Kung H-C. Final Data for 2003. *National Vital Statistics Report* 54:(13), April 19, 2006.

10. Surveillance, Epidemiology, and End Results (SEER) Program (www.seer.cancer.gov). *Mortality, Total U.S. (1969–2005).* National Cancer Institute, DCCPS, Surveillance Research Program, Cancer Statistics Branch, April 2008. Underlying mortality data provided by NCHS (www.cdc.gov/nchs).
11. Hayat MJ, Howlader N, Reichman ME, Edwards BK. Cancer statistics, trends, and multiple primary cancer analyses from the Surveillance, Epidemiology, and End Results (SEER) program. *The Oncologist* 12:20–37, 2007.
12. Lynn A. Gloeckler Ries and Milton P. Eisner. Chapter 9, Cancer of the Lung. In, Ries LAG, Young JL, Keel GE, Eisner MP, Lin YD, Horner M-J (eds). *SEER Survival Monograph: Cancer Survival Among Adults: U.S. SEER Program, 1988–2001, Patient and Tumor Characteristics.* National Cancer Institute, SEER Program, NIH Pub. No. 07-6215, Bethesda, MD, 2007.
13. SEER. *Contents of the SEER Cancer Statistics Review, 1975–2005.* http://seer.cancer.gov/csr/1975_2005/sections.html.
14. Bailar JC, Gornik HL. Cancer undefeated. *N Engl J Med* 336:1569–1574, 1997.
15. Leaf C. Why We're Losing The War On Cancer: And How To Win It. *Fortune Magazine,* March 22, 2004.
16. MedicineNet. *Better and Longer Survival for Cancer Patients.* http://www.medicinenet.com/script/main/art.asp?articlekey=157.
17. Kaiser J. NCI goal aims for cancer victory by 2015. *Science* 299:1297–1298, 2003.
18. Gemcitabine, a chemotherapeutic agent for pancreatic cancer. *NCI Cancer Bulletin* 5(12), June 10, 2008. http://www.cancer.gov/ncicancerbulletin/NCI_Cancer_Bulletin_061008/page3.
19. Kolata G. Pollack A. Costly cancer drug offers hope, but also a dilemma. *The New York Times,* July 6, 2008.
20. Yank V, Rennie D, Bero LA. Financial ties and concordance between results and conclusions in meta-analyses: retrospective cohort study. *BMJ* 335:1202–1205, 2007.
21. Sakr WA, Haas GP, Cassin BF, Pontes JE, Crissman JD. The frequency of carcinoma and intraepithelial neoplasia of the prostate in young males. *J Urol* 150:379–385, 1993.
22. Guileyardo JM, Johnson WD, Welsh RA, Akazaki K, Correa P. Prevalence of latent prostate carcinoma in two U.S. populations. *J Natl Cancer Inst* 65:311–316, 1980.
23. Gorey KM, Holowary EJ, Fehringer G, Laukkanen E, Moskowitz A, Webster DJ, Richter NL. An international comparison of cancer survival: Toronto, Ontario, and Detroit, Michigan, metropolitan areas. *Am J Public Health* 87:1156–1163, 1997.

24. Oesterling JE, Jacobsen ST, Chute CG, Guess HA, Girman CJ, Panser LA, Lieber MM. Serum prostate-specific antigen in a community-based population of healthy men. *JAMA* 270:860–864, 1993.
25. Oesterling JD, Jacobsen SJ, Cooner WH. The use of age-specific reference ranges for serum prostate specific antigen in men 60 years old or older. *J Urol* 153:1160–1163, 1995.
26. Sawyer R, Berman JJ, Borkowski A, Moore GW. Elevated prostate-specific antigen levels in black men and white men. *Modern Pathology* 9:1029–1032, 1996.
27. Oesterling JE, Kumamoto Y, Tsukamoto T, Girman CJ, Guess HA, Masumori N, Jacobsen SJ, Lieber MM. Serum prostate specific antigen in a community-based population of healthy Japanese men: lower values than for similarly aged white men. *Br J Urol* 75:347–352, 1995.
28. Luke C, Priest K, Roder D. Changes in incidence of *in situ* and invasive breast cancer by histology type following mammography screening. *Asian Pac J Cancer Prev* 7:69–74, 2006.
29. Ioannidis JP. Why most published research findings are false. *PLoS Med* 2:e124, 2005.
30. Ioannidis JP. Some main problems eroding the credibility and relevance of randomized trials. *Bull NYU Hosp Jt Dis* 66:135–139, 2008.
31. Dickersin K, Rennie D. Registering clinical trials. *JAMA* 290:51, 2003.
32. *Registering clinical trials with ClinicalTrials.gov.* http://prsinfo.clinicaltrials.gov/registering.pdf. ClinicalTrials.gov is a directory of federally and privately supported medical research trials. Section 113 of the FDA Modernization Act mandates registration with ClinicalTrials.gov of investigational new drug (IND) efficacy trials for serious diseases or conditions. The ClinicalTrials.gov Web site is a free service of the U.S. National Institutes of Health (NIH).
33. *Prostate, Lung, Colorectal and Ovarian Cancer Screening Trial (PLCO).* http://www3.cancer.gov/prevention/plco/. PLCO is an example of a large, long, and expensive clinical trial.
34. *SEER Cancer Stat Fact Sheets.* Table I-26, Age-adjusted U.S. death rates and trends for the top 15 cancer sites. http://seer.cancer.gov/csr/1975_2005/results_merged/topic_topfifteen.pdf.
35. *Funding for Various Research Areas.* http://obf.cancer.gov/financial/historical.htm.
36. Stoler DL, Chen N, Basik M, et al. The onset and extent of genomic instability in sporadic colorectal tumor progression. *Proc Natl Acad Sci USA* 96:15121–15126, 1999.
37. Holland Frei *Cancer Medicine.* Kufe D, Pollock R, Weichselbaum R, Bast R, Gansler T, Holland J, Frei E, eds. Ontario, Canada: BC Decker, 2003.

38. OMIM. Online Mendelian Inheritance in Man. http://www.ncbi.nlm.nih.gov/Omim/. OMIM is a marvelous and free database of the genetic diseases of mankind. OMIM lists every known inherited condition in man. Each condition is annotated with biologic and clinical descriptions in a detailed textual narrative that includes a listing of relevant citations. There are nearly 17,000 conditions described and OMIM.
39. PubMed home. http://www.ncbi.nlm.nih.gov/sites/entrez.
40. National Organization for Rare Diseases. http://www.rarediseases.org/.
41. Royds JA, Iacopetta B. p53 and disease: when the guardian angel fails. *Cell Death Differ* 13:1017–1026, 2006.
42. Marshall E. Genetic testing. Families sue hospital, scientist for control of Canavan gene. *Science* 290:1062, 2000.
43. Vezzosi D, Bennet A, Caron P. Recent advances in treatment of medullary thyroid carcinoma. *Ann Endocrinol* (Paris) 68:147–153, 2007.
44. Blanquet V, Turleau C, Gross-Morand MS, Senamaud-Beaufort C, Doz F, Besmond C. Spectrum of germline mutations in the RB1 gene: a study of 232 patients with hereditary and nonhereditary retinoblastoma. *Hum Molec Genet* 4:383–388, 1995.
45. Johnson RL, Rothman AL, Xie J, Goodrich LV, Bare JW, Bonifas JM, et al. Human homolog of patched, a candidate gene for the basal cell nevus syndrome. *Science* 272:1668–1671, 1996.
46. Sharp GB. The relationship between internally deposited alpha-particle radiation and subsite-specific liver cancer and liver cirrhosis: an analysis of published data. *J Radiat Res* 43:371–380, 2002.
47. Wagoner JK. Toxicity of vinyl chloride and poly(vinyl chloride): a critical review. *Environ Health Perspect* 52:61–66, 1983.
48. Hill RB, Anderson RE: Pathologists and the autopsy. *Am J Clin Pathol* 95:(Suppl)42, 1991. The authors extol the virtues of autopsies.
49. Herbst AL, Ulfelder H, Poskanzer DC. Association of maternal stilbestrol therapy and tumor appearance in young women. *New Engl J Med* 284:878–881, 1971.
50. Herbst AL, Scully RE, Robboy SJ. The significance of adenosis and clear-cell adenocarcinoma of the genital tract in young females. *J Reprod Med* 15:5–11, 1975.
51. Strohsnitter WC, Noller KL, Hoover RN, Robboy SJ, Palmer JR, et al. Cancer risk in men exposed in utero to diethylstilbestrol. *J Natl Cancer Inst* 93:545–551, 2001.
52. Wang Z, Cummins JM, Shen D, Cahill DP, Jallepalli PV, Wang TL, et al. Three classes of genes mutated in colorectal cancers with chromosomal instability. *Cancer Res* 64:2998–3001, 2004.

53. Wade N. Scientist at work: David B. Goldstein, a dissenting voice as the genome is sifted to fight disease. *The New York Times*, September 15, 2008.
54. Berman J, O'Leary TJ. Gastrointestinal stromal tumor workshop. *Hum Pathol* 32(6):578–582, 2001.
55. O'Leary T, Berman JJ. Gastrointestinal stromal tumors: answers and questions. *Hum Pathol* 33:456–458, 2002.
56. Druker BJ. Perspectives on the development of a molecularly targeted agent. *Cancer Cell* 1:31–36, 2002.
57. Virchow R. *Die Cellularpathologie in ihrer Begrundung auf physiologische und pathologische Gewebelehre.* Berlin: August Hirschwald, 1858.
58. Visscher DW, Wallis TL, Crissman JD. Evaluation of chromosome aneuploidy in tissue sections of preinvasive breast carcinomas using interphase cytogenetics. *Cancer* 77:315–320, 1996.
59. Boone CW, Bacus JW, Bacus JV, Steele VE, Kelloff GJ. Properties of intraepithelial neoplasia relevant to cancer chemoprevention and to the development of surrogate end points for clinical trials. *Proc Soc Exp Biol Med* 216:151–165, 1997.
60. Giaretti W. A model of DNA aneuploidization and evolution in colorectal cancer. *Lab Invest* 71:904–910, 1994.
61. Thomas RM, Berman JJ, Yetter, RA, Moore GW, Hutchins GM. Liver cell dysplasia: a DNA aneuploid lesion with distinct morphologic features. *Hum Pathol* 23:496–503, 1992.
62. Shin SJ, Simpson PT, Da Silva L, Jayanthan J, Reid L, Lakhani SR, Rosen PP. Molecular evidence for progression of microglandular adenosis (MGA) to invasive carcinoma. *Am J Surg Pathol* Nov 26, 2008.
63. McNeal JE, Bostwick DG. Intraductal dysplasia: a premalignant lesion of the prostate. *Hum Pathol* 17:64–71, 1986.
64. Sakr WA, Grignon DJ, Haas GP, Heilbrun LK, Pontes JE, Crissman JD. Age and racial distribution of prostatic intraepithelial neoplasia. *Eur Urol* 30:138–144, 1996.
65. Berman JJ, Henson DE. Classifying the precancers: A metadata approach. *BMC Med Inform Decis Making* 3:8, 2003.
66. Henson DE, Albores-Saavedra JA: *Pathology of Incipient Neoplasia.* New York: Oxford University Press, 2001.
67. Hruban RH, Wilentz RE, Maitra A. Identification and analysis of precursors to invasive pancreatic cancer. *Methods Mol Med* 103:1–13, 2005.
68. Brewer MA, Johnson K, Follen M, Gershenson D, Bast R Jr. Prevention of ovarian cancer: intraepithelial neoplasia. *Clin Cancer* Res 9:20–30, 2003.

69. Hui CH, Horvath N, Lewis I, To LB, Szabo F. Outcome of patients with myelodysplastic syndromes—experience from a single institution in South Australia. *Intern Med J* 38:824–828, 2008. 27% of patients with myelodysplastic syndrome progressed to acute myelogenous leukemia.
70. International Myeloma Working Group. Criteria for the classification of monoclonal gammopathies, multiple myeloma and related disorders: a report of the International Myeloma Working Group. *Br J Haematol* 121:749–757, 2003.
71. Clark WH Jr, Reimer RR, Greene M, Ainsworth AM, Mastrangelo MJ. Origin of familial malignant melanomas from heritable melanocytic lesions. "The B-K mole syndrome." *Arch Dermatol* 114:732–738, 1978.
72. Seidman JD, Berman JJ: Premalignant non-epithelial lesions: a biological classification. *Mod Pathol* 6:544–554, 1993. Though there are many published descriptions of precancerous epithelial neoplasms, there is a dearth of literature on lesions that precede the development of sarcomas and lymphomas. This paper attempts to list and describe the nonepithelial precancers.
73. Berman JJ, Albores-Saavedra J, Bostwick D, Delellis R, Eble J, Hamilton SR, Hruban RH, Mutter GL, Page D, Rohan T, Travis W, Henson DE. Precancer: A conceptual working definition. Results of a consensus conference. *Cancer Detect Prev* 30(5):387–394, 2006.
74. Piana S, Asioli S, Foroni M. Oncocytic adenocarcinoma of the rectum arising on a villous adenoma with oncocytic features. *Virchows Arch* 448:228–231, 2006.
75. Nasiell K, Nasiell M, Vaclavinkova V. Behavior of moderate cervical dysplasia during long-term follow-up. *Obstet Gynecol* 61:609–614, 1983.
76. Fu YS, Reagan JW, Richart RM. Definition of precursors. *Gynecol Oncol* 12 (Suppl):220–231, 1981.
77. Foulds L. *Neoplastic Development.* New York: Academic Press, 1969.
78. Solt D, Medline A, Farber E. Rapid emergence of carcinogen-induced initiated hepatocytes in liver carcinogenesis. *Am J Pathol* 88:595–618, 1977.
79. McDonnell TJ, Korsmeyer SJ. Progression from lymphoid hyperplasia to high-grade malignant lymphoma in mice transgenic for the t(14;18). *Nature* 349:254–256, 1991.
80. Brash DE, Ponten J. Skin precancer. *Cancer Surv* 32:69–113, 1998.
81. Berman JJ, Moore GW. The role of cell death in the growth of preneoplastic lesions: a Monte Carlo simulation model. *Cell Proliferation* 25:549–557, 1992. Regression of precancerous lesions is common. This paper examines the hypothesis that early lesions operate under the identical growth kinetics of "late" lesions (neoplasms), but that kinetic features favoring continuous growth in established lesions tend to favor extinction of lesions composed of small numbers of cells. Growth simulations of early lesions were produced using the Monte Carlo

method. The model demonstrates that small increments in the intrinsic cell loss probability in even the earliest progenitors of malignancy can strongly influence the subsequent development of neoplasia from initiated foci.

82. Berman JJ, Moore GW. Spontaneous regression of residual tumor burden: prediction by Monte Carlo Simulation. *Anal Cellul Pathol* 4:359–368, 1992. This manuscript finds a plausible explanation for the clinically observed failure of tumors to recur in instances where tumor burden remains following cancer therapy. The paper also shows that the Monte Carlo method can simulate biologic events in populations when the fate of each member of a population can be modeled probabilistically.
83. Moretti S, Spallanzani A, Pinzi C, Prignano F, Fabbri P. Fibrosis in regressing melanoma versus nonfibrosis in halo nevus upon melanocyte disappearance: could it be related to a different cytokine microenvironment? *J Cutan Pathol* 34:301–308, 2007.
84. Razack AH. Bacillus calmette-guerin and bladder cancer. *Asian J Surg* 30:302–309, 2007.
85. Evans AE, Gerson J, Schnaufer L. Spontaneous regression of neuroblastoma. *Natl Cancer Inst Monogr* 44:49–54, 1976.
86. Aterman K, Schueller EF. Maturation of neuroblastoma to ganglioneuroma. *Am J Dis Child* 120:217–222, 1970
87. Bodey B. Spontaneous regression of neoplasms: new possibilities for immunotherapy. *Expert Opinion on Biological Therapy* 2:459–476, 2002.
88. Chieco-Bianchi L, Colombatti A, Collavo D, Sendo F, Aoki T, Fischinger PJ. Tumor induction by murine sarcoma virus in AKR and C58 mice. *J Exp Med* 140:1162–1179, 1974
89. Krikorian JG, Portlock CS, Cooney P, Rosenberg SA. Spontaneous regression of non-Hodgkin's lymphoma: a report of nine cases. *Cancer* 46:2093–2099, 1980.
90. Horning SJ, Rosenberg SA. The natural history of initially untreated low-grade non-Hodgkin's lymphomas. *N Engl J Med* 311:1471–1475, 1984.
91. Tavassoli FA, Norris HJ. A comparison of the results of long-term follow-up for atypical intraductal hyperplasia and intraductal hyperplasia of the breast. *Cancer* 65:518–529, 1990.
92. Westbrook KC, Gallagher HS. Intraductal carcinoma of the breast. A comparative study. *Am J Surg* 130:667–670, 1975.
93. SEER Cancer Stat Fact Sheets. *Cancer of the breast.* http://seer.cancer.gov/statfacts/html/breast.html.
94. Chang YS, Kim Y, Kim DY, Kim HJ, Ahn CM, Lee DY, Paik HC. Two cases of post transplant lymphoproliferative disorder in lung transplant recipients. *Korean J Intern Med* 19:276–281, 2004.

95. Wallace K, Baron JA, Cole BF, et al. Effect of calcium supplementation on the risk of large bowel polyps. *J Natl Cancer Inst* 96:921–925, 2004.

96. Giardiello FM, Hamilton SR, Krush AJ, Piantadosi S, Hylind LM, Celano P, Booker SV, Robinson CR, Offerhaus GJ. Treatment of colonic and rectal adenomas with sulindac in familial adenomatous polyposis. *N Engl J Med* 328:1313–1316, 1993.

97. Berman JJ. Developmental lineage classification and taxonomy of neoplasms. http://www.julesberman.info/neoclxml.gz. This is the web location for the latest version of the Developmental Classification.

98. Reznik-Schuller H. Sequential morphologic alterations in the bronchial epithelium of Syrian golden hamsters during N-nitrosomorpholine-induced pulmonary tumorigenesis. *Am J Pathol* 89:59–66, 1977.

99. Malcovati L, Nimer SD. Myelodysplastic syndromes: diagnosis and staging. *Cancer Control* 15:4–13 (Suppl), 2008.

100. Zhao R, Xing S, Li Z, et al. Identification of an acquired JAK2 mutation in polycythemia vera. *J Biol Chem* 280:22788–22792, 2005.

101. Tefferi A, Gilliland DG. Oncogenes in myeloproliferative disorders. *Cell Cycle* 6:550–566, 2007.

102. Barosi G, Bergamaschi G, Marchetti M, Vannucchi AM, Guglielmelli P, Antonioli E, et al. JAK2 V617F mutational status predicts progression to large splenomegaly and leukemic transformation in primary myelofibrosis. *Blood* 110:4030–4036, 2007.

103. Sidon P, El Housni H, Dessars B, Heimann P. The JAK2V617F mutation is detectable at very low level in peripheral blood of healthy donors. *Leukemia* 20:1622, 2006.

104. Clarkson B, Strife A, Wisniewski D, Lambek CL, Liu C. Chronic myelogenous leukemia as a paradigm of early cancer and possible curative strategies. *Leukemia* 17:1211–1262, 2003.

105. Mukiibi JM, Nyirenda CM, Adewuyi JO, Mzula EL, Magombo ED, Mbvundula EM. Leukaemia at Queen Elizabeth Central Hospital in Blantyre, Malawi. *East Afr Med* J 78:349–354, 2001.

106. Bacher U, Haferlach T, Hiddemann W, Schnittger S, Kern W, Schoch C. Additional clonal abnormalities in Philadelphia-positive ALL and CML demonstrate a different cytogenetic pattern at diagnosis and follow different pathways at progression. *Cancer Genet Cytogenet* 157:53–61, 2005.

107. Lane CM, Guo XY, Macaluso LH, Yung KC, Deisseroth AB. Presence of P210bcrabl is associated with decreased expression of a beta chemokine C10 gene in a P210bcrabl-positive myeloid leukemia cell line. *Mol Med* 5:55–61, 1999.

108. Skorski T. BCR/ABL, DNA damage and DNA repair: implications for new treatment concepts. *Leuk Lymphoma* 49:610–614, 2008.

109. Calabrese P, Tavare S, Shibata D. Pretumor progression: clonal evolution of human stem cell populations. *Am J Pathol* 164:1337–1346, 2004.

110. Berman JJ. *Neoplasms: Principles of Development and Diversity.* Sudbury, MA: Jones & Bartlett, 2009.

111. Beckwith JB, Kiviat NB, Bonadio JF. Nephrogenic rests, nephroblastomatosis, and the pathogenesis of Wilms' tumour. *Pediatr Pathol* 10:1–36, 1990.

112. Park S, Bernard A, Bove KE, Sens DA, Hazen-Martin DJ, Garvin AJ, Haber DA. Inactivation of WT1 in nephrogenic rests, genetic precursors to Wilms' tumour. *Nat Genet* 5:363–367, 1993.

113. Beckwith JB, Perrin EV. In situ neuroblastomas: a contribution to the natural history of neural crest tumors. *Am J Pathol* 43:1089–1104, 1963. *In situ* neuroblastomas are small adrenal tumors, found incidentally in some neonatal autopsies. They are cytologically identical to neuroblastomas. Because they occur at a higher incidence than neuroblastomas, we infer that these lesions often regress.

114. Bessho F. Effects of mass screening on age-specific incidence of neuroblastoma. *Int J Cancer* 67:520–522, 1996.

115. Komoto M, Tominaga K, Nakata B, Takashima T, Inoue T, Hirakawa K. Complete regression of low-grade mucosa-associated lymphoid tissue (MALT) lymphoma in the gastric stump after eradication of *Helicobacter pylori. J Exp Clin Cancer Res* 25:283–285, 2006.

116. Berman JJ, Henson DE. Reply. *Hum Pathol* 35:137–138, 2004.

117. Berman JJ, Henson DE. The precancers: Waiting for a classification. *Human Pathol* 34:833–834, 2003.

118. NIOSH carcinogen list. http://www.cdc.gov/niosh/npotocca.html. The list of substances that NIOSH considers to be potential occupational carcinogens is actually quite short, despite the often-repeated complaint that "everything is a carcinogen" in animal tests.

119. Cranor C. Scientific inferences in the laboratory and the law. *Am J Public Health* 95:S121–S128, 2005.

120. National Academy of Sciences. *Toxicity Testing: Strategies to Determine Needs and Priorities.* Washington, DC: National Academy Press, 1984.

121. National Toxicology Program. *11th Report on Carcinogens.* http://ntp.niehs.nih.gov/index.cfm?objectid=72016262-BDB7-CEBA-FA60E922B18C2540.

122. Bradford HA. *Principles of Medical Statistics.* London: The Lancet, 1971.

123. Rothman KJ. *Modern Epidemiology.* Boston: Little, Brown, 1986.

124. Huff JE. Value, validity, and historical development of carcinogenesis studies for predicting and confirming carcinogenic risks to humans. In: Kitchin KT, ed. *Carcinogenicity Testing, Predicting, and Interpreting Chemical Effects.* New York: Marcel Dekker, 1999:21–123.

125. Food and Drug Administration's Modernization Act. http://www.fda.gov/cder/guidance/105-115.htm.

126. Kelloff GJ, O'Shaughnessy JA, Gordon GB, et al. Counterpoint: Because some surrogate end point biomarkers measure the neoplastic process they will have high utility in the development of cancer chemopreventive agents against sporadic cancers. *Cancer Epidemiol Biomarkers Prev* 12:593–596, 2003.

127. Kelloff GJ. Sigman CC, Johnson KM, Boone CW, Greenwald P, Crowell JA, Hawk ET, Doody LA. Perspectives on surrogate end points in the development of drugs that reduce the risk of cancer. *Cancer Epidemiol Biomarkers Prev* 9:127–37, 2000.

128. Wargovich MJ, Jimenez A, McKee K, et al. Efficacy of potential chemopreventive agents on rat colon aberrant crypt formation and progression. *Carcinogenesis* 21:1149–1155, 2000.

129. Williams D, Verghese M, Walker LT, Boateng J, Shackelford L, Chawan CB. Flax seed oil and flax seed meal reduce the formation of aberrant crypt foci (ACF) in azoxymethane-induced colon cancer in Fisher 344 male rats. *Food Chem Toxicol* 45:153–159, 2007.

130. Krizman DB, Chuaqui RF, Meltzer PS, et al. Construction of a representative cDNA library from prostatic intraepithelial neoplasia. *Cancer Res* 56:5380–5383, 1996.

131. Biology of Breast Pre-Malignancies (R01). http://grants.nih.gov/grants/guide/rfa-files/RFA-CA-07-047.html.

132. Kochanek KD, Murphy SL, Anderson RN, Scott C. Deaths: Final data for 2002. *National Vital Statistics Report* 53:(5), October 12, 2004. http://www.cdc.gov/nchs/data/nvsr/nvsr53/nvsr53_05.pdf. Contains a data summary for the leading causes of death in the United States.

133. Tsai SP, Lee ES, Hardy RJ. The effect of a reduction in leading causes of death: potential gains in life expectancy. *Am J Public Health* 68:966–971, 1978.

134. Central Intelligence Agency. *World Factbook.* Rank-order life expectancy at birth. https://www.cia.gov/library/publications/the-world-factbook/rankorder/2102rank.html.

135. Puska P. Successful prevention of non-communicable diseases: 25 year experiences with North Karelia Project in Finland. *Public Health Medicine* 4:5–7, 2002.

136. Wabinga HR: Pattern of cancer in Mbarara, Uganda. *East Afr Med J* 79:193–197, 2002.

137. Nze-Nguema F, Sankaranarayanan R, Barthelemy M, et al. Cancer in Gabon, 1984–1993: A pathology registry-based relative frequency study. *Bull Cancer* 83:693–696, 1996.

138. Palatianos GM, Cintron JR, Narula T, et al: Father of modern cytology. A 30-year commemorative. *J Fla Med Assoc* 79:837–838, 1992.

139. Bergstrom R, Sparen P, Adami HO: Trends in cancer of the cervix uteri in Sweden following cytological screening. *Br J Cancer* 81:159–166, 1999.

140. Anttila A, Pukkala E, Soderman B, et al: Effect of organised screening on cervical cancer incidence and mortality in Finland, 1963–1995: Recent increase in cervical cancer incidence. *Int J Cancer* 83:59–65, 1999.

141. Alberts DS. Reducing the risk of colorectal cancer by intervening in the process of carcinogenesis: a status report. *Cancer J* 8:208–221, 2002.

142. Fisher B, Costantino JP, Wickerham DL, et al. Tamoxifen for prevention of breast cancer: report of the National Surgical Adjuvant Breast and Bowel Project P-1 Study. *J Natl Cancer Inst* 90:1371–1388, 1998.

143. Kolata G. Reversing trend, big drop is seen in breast cancer. *The New York Times*, December 15, 2006.

144. Correa P, Fontham ET, Bravo JC, Bravo LE, Ruiz B, Zarama G, et al. Chemoprevention of gastric dysplasia: randomized trial of antioxidant supplements. *J Natl Cancer Inst* 92:1881–1888, 2000.

145. Orbo A, Rise CE, Mutter GL. Regression of latent endometrial precancers by progestin infiltrated intrauterine device. *Cancer Res* 66:5613–5617, 2006.

146. Saba HI. Decitabine in the treatment of myelodysplastic syndromes. *Ther Clin Risk Manag* 3:807–817, 2007. Decitabine is a hypomethylating agent that promotes the reexpression of tumor suppressor genes. It has a beneficial effect on patients with myelodysplasia, slowing the progression of the disease to acute myeloid leukemia.

147. Finn OJ. Premalignant lesions as targets for cancer vaccines. *J Exp Med* 198:1623–1626, 2003.

148. Ackerman AB. Request, respectfully, return to rudiments. *Hum Pathol* 351:136–137, 2004.

149. O'Shaughnessy JA, Kelloff GJ, Gordon GB, Dannenberg AJ, Hong WK, Fabian CJ, Sigman CC, Bertagnolli MM, Stratton SP, Lam S, et al. Recommendations of the American Association for Cancer Research Task Force on the Treatment and Prevention of Intraepithelial Neoplasia. Treatment and prevention of intraepithelial neoplasia: An important target for accelerated new agent development. *Clin Cancer Res* 8:314–346, 2002.

150. Frey CM, McMillen MM, Cowan CD, Horm JW, Kessler LG. Representativeness of the surveillance, epidemiology, and end results program data: recent

trends in cancer mortality rate. *JNCI* 84:872, 1992. The Surveillance, Epidemiology, and End Results (SEER) project collects cancer-related incidence and mortality data collected from residents in geographically defined populations, representing about 10% of the U.S. population.

151. Ashworth TG. Inadequacy of death certification: proposal for change. *J Clin Pathol* 44:265, 1991. A British perspective on the importance of the death certificate.
152. Kircher T, Anderson RE. Cause of death: proper completion of the death certificate. *JAMA* 258:349–352, 1987. Though every physician is expected to complete death certificates, surprisingly few physicians understand how to do the job. As a consequence, death certificates are notoriously inadequate records of the cause of death. The authors explain the differences between the underlying and immediate causes of death and the mechanism and manner of death.
153. Walter SD, Birnie SE. Mapping mortality and morbidity patterns: an international comparison. *Intl J Epidemiol* 20:678–689, 1991. This survey of 49 national and international health atlases has shown that there is virtually no consistency in the way that death data are presented.
154. Demographia. *U.S. population from 1900.* http://www.demographia.com/db-uspop1900.htm.
155. *SEER Cancer Statistics Review 1975–2005.* Table II-2. All cancer sites (invasive) age adjusted seer incidence rates by year, race and sex. http://seer.cancer.gov/csr/1975_2005/results_merged/topic_annualrates.pdf.
156. Central Intelligence Agency. *The World Factbook.* https://www.cia.gov/library/publications/the-world-factbook/.
157. Shibuya K, Mathers CD, Boschi-Pinto C, Lopez AD, Murray CJ. Global and regional estimates of cancer mortality and incidence by site: II. Results for the global burden of disease 2000. *BMC Cancer* 2:37, 2002.
158. SEER Stat Fact Sheets. http://seer.cancer.gov/statfacts/html/all.html.
159. Tzen C, Huang Y, Fu Y. Is atypical follicular adenoma of the thyroid a preinvasive malignancy? *Hum Pathol* 34:666–669, 2003.
160. Schnitt SJ. The diagnosis and management of pre-invasive breast disease: flat epithelial atypia—classification, pathologic features and clinical significance. *Breast Cancer Res* 5:263–268, 2003.
161. Guitart J, Magro C. Cutaneous T-cell lymphoid dyscrasia: a unifying term for idiopathic chronic dermatoses with persistent T-cell clones. *Arch Dermatol* 143:921–932, 2007.
162. Greenberg AK, Yee H, Rom WN. Preneoplastic lesions of the lung. *Respir Res* 3:20–30, 2002.

163. Gurney JG, Davis S, Severson RK, Fang JY, Ross JA, Robison LL.Trends in cancer incidence among children in the U.S. *Cancer* 78:532–541, 1996.
164. SEER Fast Stats. Statistics Stratified by Data Type. http://seer.cancer.gov/faststats/selections.php.
165. Khurana V, Bejjanki HR, Caldito G, Owens MW. Statins reduce the risk of lung cancer in humans: a large case-control study of US veterans. *Chest* 131:1282–1288, 2007.
166. Jemal A, Murray T, Ward E, et al. Cancer statistics, 2005. *CA Cancer J Clin* 55:10–30, 2005.
167. Werb P, Scurry J, Oumlstoumlr A, Fortune D, Attwood H. Survey of congenital tumors in perinatal necropsies. *Pathology* 24:247–253, 1992.
168. Mutter GL. Diagnosis of premalignant endometrial disease. *J Clin Pathol* 55:326–331, 2002.
169. Rosenwald A, Wright G, Leroy K, Yu X, Gaulard P, Gascoyne RD, et al. Molecular diagnosis of primary mediastinal B cell lymphoma identifies a clinically favorable subgroup of diffuse large B cell lymphoma related to Hodgkin lymphoma. *J Exp Med* 198:851–862, 2003.
170. Schuetz AN, Yin-Goen Q, Amin MB, Moreno CS, Cohen C, Hornsby CD, et al. Molecular classification of renal tumors by gene expression profiling. *J Mol Diagn* 7:206–218, 2005.
171. Zhou X, Temam S, Oh M, Pungpravat N, Huang BL, Mao L, Wong DT. Global expression-based classification of lymph node metastasis and extracapsular spread of oral tongue squamous cell carcinoma. *Neoplasia* 8:925–932, 2006.
172. U.S. Orphan Drug Act of 1983. http://www.fda.gov/orphan/oda.htm. Before the passage of this act, there was virtually no development of new drugs for the treatment of rare diseases. The Orphan Drug Act of 1983 encouraged the development of new drugs for rare diseases by offering tax advantages and an extended period of exclusive marketing rights for participating companies.
173. Tatsioni A, Bonitsis NG, Ioannidis JP. Persistence of contradicted claims in the literature. *JAMA* 298(21):2517–2526, 2007.

Index

A

D

E

F

G

K

L

M

N

O

P

R

S

T

U

V

W

X

Z